Second Edition

Optical Imaging
Techniques
in
Cell Biology

Second Edition

Optical Imaging Techniques in Cell Biology

Guy Cox

Australian Centre for Microscopy and Microanalysis
University of Sydney

CRC Press
Taylor & Francis Group
Boca Raton London New York

CRC Press is an imprint of the
Taylor & Francis Group, an **informa** business

CRC Press
Taylor & Francis Group
6000 Broken Sound Parkway NW, Suite 300
Boca Raton, FL 33487-2742

First issued in paperback 2017

ISBN-13: 978-1-4398-4825-8 (hbk)
ISBN-13: 978-1-138-19940-8 (pbk)

Library of Congress Cataloging-in-Publication Data

Optical imaging techniques in cell biology / editor, Guy Cox. -- 2nd ed.
    p. ; cm.
  Rev. ed. of: Optical imaging techniques in cell biology / Guy Cox. c2007.
  Includes bibliographical references and index.
  ISBN 978-1-4398-4825-8 (hardcover : alk. paper)
  I. Cox, Guy, 1945- II. Cox, Guy, 1945- Optical imaging techniques in cell biology.
  [DNLM: 1. Cytological Techniques--methods. 2. Microscopy--methods. QY 95]

  571.6--dc23   2012013732

Visit the Taylor & Francis Web site at
http://www.taylorandfrancis.com

and the CRC Press Web site at
http://www.crcpress.com

Contents

Introduction: The Optical Microscope in Cell Biology

The simplest microscope is a single lens. The word "lens" is Latin, and lenses seem to have been known to the ancient Greeks and Romans, as well as to the Chinese, for thousands of years. Lenses were used as *burning glasses* to start fires by focusing the light of the sun on kindling, and in medieval Europe (and much earlier in China) they were used to correct vision. Yet the idea of using them to look at small objects does not seem to have occurred to the ancients. The problem was philosophical, not technical—they simply had no concept that there was anything significant to see beyond the limits of the human eye. The world, to them, was what they could see.

It was not until 1656 (according to the *Oxford English Dictionary*) that the word *microscope* entered the English language, and its equivalent is recorded in Italian at around the same time. This is far more significant than the coining of a new word; it reflects the new way of looking at the world that emerged in the Renaissance and grew into the Age of Enlightenment. There was a whole cosmos to study with telescopes—a universe hugely larger than that imagined by the ancients. And then there was the microcosm—the world below the scale of human sight, studied with the microscope.

A book was published in 1665 that changed the way people looked at the world. It was the first, and most famous, book on microscopy ever written: Robert Hooke's *Micrographia*, published by the Royal Society. Hooke was not confident enough to use the new word *microscope* on the title page, referring instead to "physiological descriptions of minute bodies made by magnifying glasses." But what he used was a compound microscope, with a focusing system, illuminator, and condenser, all recognizable albeit primitive. It was not very good, but one of his observations had great significance—he called the compartments he could see in cork "cells." If fact, cells are also clearly shown in his drawing of a nettle leaf, though he did not comment on them in the text.

Then, and for at least two hundred years afterward, the *simple microscope* consisting of a single lens was also the best microscope, able to give better resolution than *compound microscopes* using multiple lenses such as Hooke's. With the aid of very carefully made single lens microscopes, Hooke's contemporary, the Dutch naturalist Antonie van Leeuwenhoek, was able to take his microscopic studies to a level of resolution an order of magnitude better than Hooke's. van Leeuwenhoek saw and described spermatozoa, red blood corpuscles, and bacteria more than 300 years ago. (He took pains to explain that the sperm "were obtained from the overplus of marital intercourse and without committing any sin".) It is impossible to overestimate the significance of these discoveries. Imagine a world in which nobody knew what caused disease, or how conception worked, or what blood did. Van Leeuwenhoek did not answer these questions, but he made them answerable.

Compound microscopes were used for greater convenience and versatility, but not until the late 19th century could they surpass simple microscopes in performance. They were inferior because of the aberrations (defects) of the lenses. The more lenses that were added to a microscope the worse it got because each one magnified the level of aberration. This topic is dealt with in detail in Chapter 7.

The last person to make major scientific discoveries with the simple single-lens microscope was the legendary botanist Robert Brown in the early 19th century. His name is perpetuated in Brownian motion, which he discovered (and understood), and he also studied fertilization in plants, describing the pollen tube and its function. In the course of this study—almost in passing—he described the cell nucleus. It was a tiny part of a very long paper, but his contemporaries spotted it and took note.

Between van Leeuwenhoek and Brown lay a timespan of over 100 years, and during that time compound microscopes were mostly the playthings of wealthy amateurs who used them mostly as expensive toys. Their quality simply was not good enough for science. This does not mean that the problems were being ignored.

In the 1730s Chester More Hall, an English barrister, set out to construct an achromatic lens, free from the color fringing that ruined the performance of current lenses. He realized that what was needed were two media of different dispersion, and fortunately for him a new glass ("flint glass," with a high lead content) had just been produced with a higher refractive index and dispersion than traditional "crown" glass. He would need to combine two lenses: a strong positive one of crown glass and a weak negative one of flint glass. To test his idea he needed to get these lenses made, but he was afraid that whoever made them for him would pinch the idea. So he commissioned two different opticians to make the two lenses. Unfortunately, both opticians subcontracted the job to the same lens maker, George Bass. Bass and Hall made several achromatic telescopes for Hall's use, the first in 1733, but neither commercialized the invention. The telescopes remained in existence, though, so Hall's priority was clearly proven.

Bass kept quiet about Hall's achromatic telescope lenses for 20 years, but then he discovered that the microscope and telescope maker John Dollond was experimenting along similar lines. Bass told Dollond about the lenses he had built for Hall, and Dollond promptly patented the idea and started production. Other optical manufacturers naturally disputed the patent. The court ruled that although Hall's priority was clear, Dollond deserved patent protection for his work in bringing the lenses to production and making them known to the world, so the patent was upheld. The company, now called Dollond & Aitchison, is still in business. From about 1758 achromatic objectives for telescopes, newly designed by Dollond but based on Hall's principles, were generally available. However, attempts to make such lenses for microscopes proved too technically difficult at the time, and it was not until the 19th century that achromatic microscope objectives became available.

The other major optical problem was spherical aberration. This can be minimized in a single lens by making it a meniscus shape. Van Leeuwenhoek discovered this empirically and Hall's first achromats also deliberately used this design to minimize spherical aberration. However, although this partial correction is not too bad in a telescope, it is not adequate in a microscope working at high magnification.

Joseph Jackson Lister, father of the surgeon who introduced antiseptic surgery, set out to solve the problem in the early 19th century. He was a wine merchant and a Quaker (an odd combination) and this was only a hobby, but his science was impeccable, and his design principles are still the starting point for all microscope objectives. His first corrected microscope was built for him in 1826, and his paper was published in the *Philosophical Transactions of the Royal Society* in 1830. Optical manufacturers did not immediately beat a path to his door, but with some persuasion Lister got the large optical firm of Andrew Ross to take over the manufacture of his corrected lenses. Ross also designed a much more rigid microscope so that the high resolution these lenses produced could be used in practice.

Lister and Ross took the design one stage further by introducing a correction collar to adjust the spherical aberration correction for use with or without a cover slip, or with different thicknesses of coverslip, and this was published in 1837. It became a must-have feature so some less reputable manufacturers provided a collar that did not actually do anything, or even just engraved markings without anything that turned. The compound microscope was now a fit tool for scientists, and Lister himself was one of the first to put it to work. He showed that red blood cells (first seen by van Leeuwenhoek) were biconcave disks and discovered that all muscle was made up of fibers.

Two German scientists, Matthias Schleiden and Theodor Schwann, also eagerly adopted the new microscopes. In 1838 Schlieden (then professor of botany at the University of Jena) wrote *Contributions to Phytogenesis*, in which he stated that the different parts of the plant organism are composed of cells or derivatives of cells. He thereby became the first to formulate what was then an informal belief as a principle of biology equal in importance to the atomic theory of chemistry. He also recognized the importance of the cell nucleus, discovered in 1831 by the Scottish botanist Robert Brown, and sensed its connection with cell division.

Schwann studied medicine at the University of Berlin, and while there at the age of 26, discovered the digestive enzyme pepsin, the first enzyme ever isolated from animal tissue. Schwann moved to Belgium, becoming professor first at Louvain (1838) then at Liège (1848). He identified striated muscle and discovered the myelin sheath of nerve cells (Schwann cells are named after him). He also showed that fermentation was the product of a living organism. Schwann's *Mikroskopische Untersuchungen über die Übereinstimmung in der Struktur und dem Wachstume der Tiere und Pflanzen* (1839; Microscopical Researches into the Accordance in the Structure and Growth of Animals and Plants) extended his friend Schleiden's work into a general theory applying to plants, animals, and protists.

The framework for cell biology was now in place, but the optics of the microscope were still not fully understood, and many microscopists were frustrated at not being able to obtain the results they thought they should get. Physicists were working on the problem. George Airy was an astronomer (in due course Astronomer Royal) who showed that the image of a point source formed by a lens of finite diameter was a disk with halos around it whose properties depended entirely on the size of the lens. This was a key point in understanding the performance of a microscope, but Airy was perhaps not the man to take it further. He was a vain and highly opinionated man who refused to accept such major breakthroughs as the discovery of Neptune and Faraday's electrical theory.

John Strutt, Lord Rayleigh, was the man who saw the use of Airy's discovery. He was a brilliant scientist who first made his name in acoustics but then moved into optics and explained how the wave nature of light determined how it was scattered (Rayleigh scattering). He also discovered the element argon. In microscopy he gave the first mathematical analysis of resolution, defining a resolution criterion based on the Airy disk and showing how it was determined by the numerical aperture of the objective.

The man who more than any other completed the evolution of the microscope was Ernst Abbe, a junior professor of physics at the University of Jena, who joined the optical company founded by the university's instrument maker Carl Zeiss in 1866, became a partner in 1876, and took over the firm after Zeiss' death in 1888. Abbe realized that Rayleigh's treatment was not correct for the common case of a specimen illuminated by transmitted or reflected light, and developed his diffraction theory of image formation, which for the first time made the significance of the illumination system clear. He also designed a more perfectly corrected lens than any before, allowing microscopes for the first time to actually reach the theoretical limit imposed by the wavelength of light.

At this time, the end of the 19th century, the microscope had reached its limit. To some, this was the end of the line but with hindsight it was a boon, for it ended the quest for more resolution and set scientists on the road to expanding the capabilities of the microscope. This was the golden age of the histologist; microtomes, too, had been perfected, a wide range of fixatives and stains was in use, and superb images were obtainable of fixed, stained sections. However, cell physiologists wanted to understand the workings of the cell, and being restricted to dead, sliced material was a serious limitation.

The 20th century brought a revolution in what could be studied in the microscope. Fluorescence came early, introduced by August Köhler, the same man who gave us Köhler illumination, though it was another 60 years before the improvements introduced by the Dutchman J.S. Ploem made it widely popular. Zernike's phase contrast (yet another Dutch invention), followed by Nomarski's differential interference contrast, gave us for the first time convenient, effective ways to study living cells. Polarized light could reveal structural information about cell components that lay below the resolution limit. Midway through the century, the optical microscope had become a serious tool for the cell biologist. Then came confocal microscopy, and multiphoton, and a huge expansion in the realm of optical imaging tools in biology.

But they do not belong in the introduction; these techniques are what this book is about.

Contributors

Teresa Dibbayawan
Electron Microscope Unit
University of Sydney
Sydney, Australia

Eleanor Kable
Electron Microscope Unit
University of Sydney
Sydney, Australia

Nuno Moreno
Cell Imaging Unit
Instituto Gulbenkian de Ciência
Oeiras, Portugal

Acknowledgments

Many people have contributed in one way or another to this book. Teresa Dibbayawan (who is also my wife) and Eleanor Kable have contributed in many ways beyond the sections with their names. Scott Kable, Eleanor's husband, is my major fount of knowledge on the physical chemistry of fluorescence and the ultimate source of many of the Jablonski diagrams. Nuno Moreno played a major part of the original genesis of this book. Colin Sheppard has been my long-term collaborator and optics guru, and his influence is large on all the advanced optics sections. My colleagues Anya Salih, Filip Braet, Lilian Soon, Allan Jones, and Louise Cole have all helped in many different ways. And I thank all the named contributors of pictures, many of whom responded rapidly and with huge enthusiasm when asked to help. All the manufacturers I contacted were extremely helpful; they are too many to list individually but I have to particularly thank Mike Stanley of Chroma Corp., Carola Thoni of Leica Microsystems, Chris Johnson of Perkin-Elmer, and Gavin Symonds of Zeiss.

On a personal level I have to thank my late parents for encouraging my interest in the natural world and its smaller aspects; and the late Sidney Chillingworth, a family friend who gave a 10-year-old boy (me) the magnificent Zeiss "Jug-Handle" microscope seen in Figure 1.5. Finally, my DPhil supervisor, Barrie Juniper, whose encouragement played a large part in making the microscopic side of biology my life's work; and the late David Cockayne, director of the Australian Key Centre for Microscopy and Microanalysis in Sydney for 26 years and my friend and colleague for much longer.

1 The Light Microscope

The simplest microscope is a single lens, and as we saw in the Introduction this was for many years the best microscope. *Simple microscopes*, as they are called, can give high-resolution images, and once we understand how one lens works, it is equally easy to see how they are combined to form both ancient and modern *compound microscopes*.

LENSES AND MICROSCOPES

There is one fundamental property of a lens from which many of its other characteristics follow on automatically. A lens, by definition, makes parallel rays of light meet at a point, called the *focus* of the lens (Figure 1.1). The distance from the lens to the focus is called the *focal length*. For a simple understanding of the microscope the focal length is the only thing we need to know about a lens. This ignores any imperfections of real lenses, assuming that all lenses are perfect, but it is a great starting point.

We can use this property to draw some very simple ray diagrams, which will show us how and where a lens will form an image and from there how different lenses work together in the complete microscope. Since all parallel rays are deflected through the focus, while rays that pass through the exact center of the lens will go straight through, undeflected, it is easy to plot where the image of a specimen will be.

Provided that the object is farther away from the lens than its focal length, we can draw these two rays, as shown in Figure 1.2, from any point on the object. Where they meet is where the image of that point will be. (If the sample is closer to the lens than its focal length, the two rays will not meet.) This shows us that the objective lens forms an *inverted, real* image—real in this context meaning that it is an actual image that can be projected on a screen. In fact you can easily do just that if you take out an eyepiece from a microscope in a darkened room. Put a piece of tracing paper over the hole and you will see this image, since it is formed at a point near the top of the microscope tube. A slight adjustment of the fine focus should bring it into sharp focus.

How then can we get a greater magnification? As Figure 1.3 shows, this is done by using a lens with a shorter focal length. A "higher power" lens really means one that has a shorter focal length. If we are going to still form the image at the same place we must bring the specimen closer to the lens, which is why high power objectives are always longer than low power ones. They are designed to be *parfocal*—as you rotate the turret of the microscope each lens should come into approximate focus once one has been focused.

In theory (and indeed in practice) we could make the image larger with any given focal length lens by forming the image farther away (i.e., bringing the specimen closer to the lens). But lenses are corrected for one image distance, as explained in Chapter 7, and the image quality will suffer if we do that. The distance from the back of the

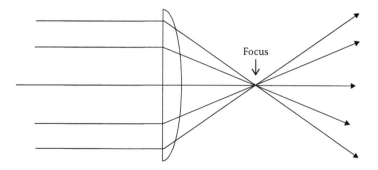

FIGURE 1.1   A lens makes parallel rays meet at a point, called the focus of the lens.

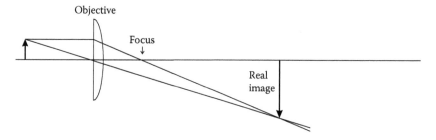

FIGURE 1.2   Plotting the paths of the parallel and central rays shows where a lens of given focal length will form an image and what the magnification will be.

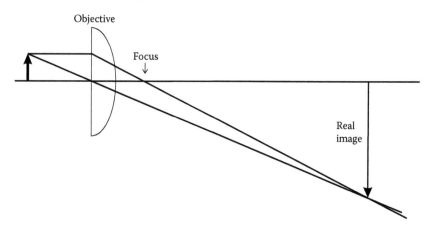

FIGURE 1.3   A high power objective has a shorter focal length. It is positioned closer to the specimen and will then form a more highly magnified image but at the same place.

objective to the image plane is called the *tube length* and is marked on the lens. Having a fixed tube length brings us the added benefit that we know what magnification an objective of a given focal length will give, and we can mark this on the lens. If you look at an old microscope objective, from the days when tube lengths were not standardized, you will find that it gives the focal length rather than the magnification.

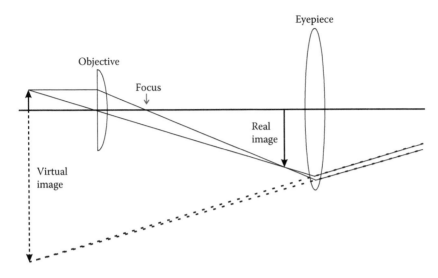

FIGURE 1.4 The eyepiece is positioned close to the real image formed by the objective. It cannot therefore form a real image and instead the rays appear to originate from a magnified virtual image.

We do not usually view the image on a sheet of tracing paper but through another lens, the *eyepiece*. The eyepiece is positioned so that the real image projected by the objective is closer to the lens than its focal length. This means, as we have seen, that it cannot form a real image. However, if we plot the rays in the same way (Figure 1.4) we see that they appear to come from a point behind the lens, which is thus a *virtual* image. The virtual image is the same way up as the original real image but much larger, so the final image remains inverted relative to the specimen but is magnified further by the eyepiece. We normally focus a microscope so that the virtual image is located somewhere around the near point of our eyes, but this is not essential. Where it forms is under our control as we adjust the focus.

Figure 1.5 shows these same ray paths in their correct position relative to an actual microscope. For the sake of simplicity an old single-eyepiece microscope is used since its optical path is in a straight line. This also emphasizes the point that microscope resolution has not improved since the end of the 19th century; the modern microscope is not really any better optically.

Each operator may focus the microscope slightly differently, forming the virtual image at a different plane. This does not cause any problems in visual microscopy, but once we start to take photographs it can make life difficult. The camera needs an image at the plane of the film, not the plane at which an individual operator finds comfortable for viewing. A microscope fitted with a camera will also have crosshairs visible through the eyepiece, and the eyepiece will have a focusing adjustment to bring these into focus. Then the microscope focus is also adjusted so that the image is seen in sharp focus with the crosshairs superimposed; it will then also be in focus for the camera.

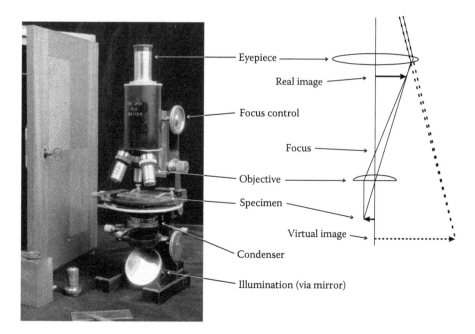

Eyepiece

Real image

Focus control

Focus

Objective

Specimen

Virtual image

Condenser

Illumination (via mirror)

FIGURE 1.5 An early 20th century Zeiss "jug-handle" microscope. By this stage the microscope had developed to the point where its resolution was limited only by the wavelength of light, and it has therefore not improved since. Modern improvements are fundamentally matters of convenience (see Figure 1.9).

The final magnification that we see will be the objective magnification multiplied by the eyepiece magnification, provided that the tube length is correct. The image captured by a camera will have a different value, determined by the camera adaptor.

Modern microscopes rarely just have the one eyepiece, which sufficed up until the mid-20th century. Binocular eyepieces just use a beamsplitter to duplicate the same image for both eyes; they do not give a stereoscopic view (Figure 1.6). (In this they differ from the beautiful two-eyepiece brass microscopes of the 19th century, which were stereoscopic but sacrificed some image quality by looking through the objective at an angle.)

There always needs to be an adjustment for the interocular spacing to suit different faces. This has to be done without altering the tube length. Sometimes there is a pivoting arrangement so that the actual length does not change. On other microscopes the eyepieces move out as they come closer together and in as they move farther apart. On some older microscopes it must be manually adjusted; the focusing ring on each eyepiece must be adjusted to match the scale reading on the interocular adjustment.

THE BACK FOCAL PLANE OF A LENS

To understand the practical side of setting up a real microscope we need to know more than just where the image is formed. We also need to look at what the light does as it passes down the tube between the objective lens and the image it forms.

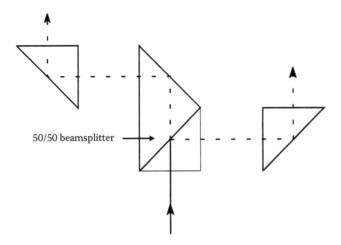

FIGURE 1.6 Typical arrangement of prisms in a binocular microscope. Both eyes see the same image.

If we construct ray paths through a lens as before but for two points on the sample equidistant from the optic axis, we have Figure 1.7A. From each point we have drawn one ray from the object passing undeflected through the center of the lens and a second from the same point running parallel to the optic axis, and therefore deflected through the focus of the lens.

With these two rays from each point we have defined the position and size of the image. Since we know that all rays from each point must arrive at the corresponding point on the image, we can now draw in as many more rays as we like. Figure 1.7B adds the rays that leave the object at an angle, α, equal to that of the rays passing through the center of the lens but in the opposite direction. An interesting feature of the ray paths is now apparent. Behind the lens, at the plane of the focus, rays leaving both object points at an angle of $+\alpha$ cross each other above the *optic axis* (the line through the center of the lens). The parallel rays, of course, cross at the focus, while at an equal distance below the optic axis the rays leaving both object points at $-\alpha$ cross each other.

More points and more rays can be added to the diagram. Figure 1.8A shows rays leaving at smaller angles, which cross at the same plane but closer to the optic axis. You might like to plot rays at larger angles and observe where they meet; plotting it yourself will explain more than any number of words. Figure 1.8B shows two more points, with rays leaving each at $+\alpha$ and $-\alpha$. The construction lines have been omitted for the sake of clarity; in any case having already defined two points on our image, we know where other object points will be imaged. What we see is that the $+\alpha$ and $-\alpha$ rays all cross at the same point.

These constructions show us that at the plane of the focus, all rays, from any point of the object, that leave in one particular angle and direction will meet at one point. The larger the angle, the further this point will be from the optic axis. This plane is called the *back focal plane* (or sometimes just focal plane) of the lens. It is also sometimes called the *pupil*. The back focal plane will lie between 1 and 10 mm behind the

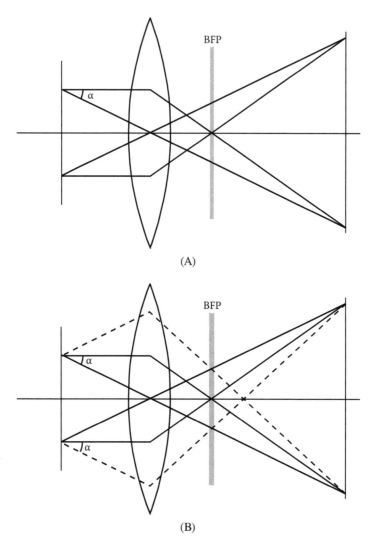

(A)

(B)

FIGURE 1.7   (A) Rays from two points on the sample to corresponding points on the image. (B) Two additional rays at matching angles added. Rays at corresponding angles cross at the same plane as the focus (the back focal plane or BFP) but off the optic axis.

lens, since this is the range of focal lengths of common microscope objectives. It will therefore be within the lens mount.

The back focal plane of the objective is easy enough to see. Simply focus the microscope on a specimen, then remove an eyepiece and look down the tube. A larger and clearer image can be obtained by using a phase telescope instead of the standard eyepiece. This is a long-focal-length eyepiece, which takes the back focal plane as its object and gives a magnified virtual image of it, just as a standard eyepiece does with the image. Alternatively, some of the larger and more elaborate research microscopes have a Bertrand lens built into the column (Figure 1.9). This is

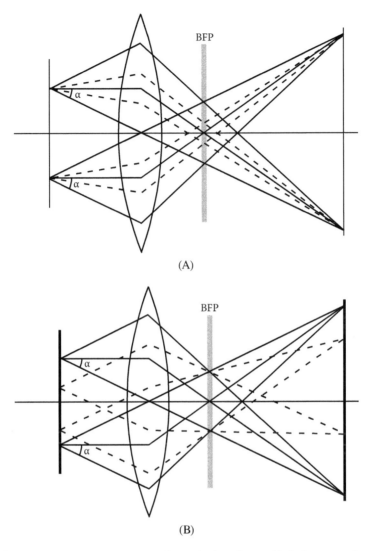

FIGURE 1.8 (A) Additional rays at smaller angles have been added; they cross closer to the optic axis. (B) Rays from all points on the image at a given angle cross at the same point on the back focal plane.

an extra lens, which can be put into position with a slider, and forms an image of the back focal plant at the normal image position, where it is magnified by the normal eyepiece just as if it were the image of the specimen.

If a specimen scatters light uniformly, the back focal plane will be a uniformly bright disc, its diameter being determined by the maximum value of α that the lens can accept. If the specimen has a tendency to scatter light in particular directions we will see this as a pattern in the back focal plane (Figure 1.9C). At the back focal plane, light can be sorted not according to where it comes from but according to the angle at which it has been scattered by the specimen. This gives a handle on controlling

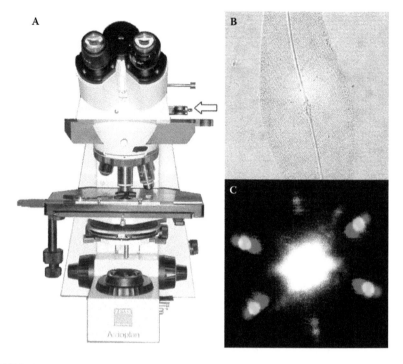

FIGURE 1.9 (A) A modern Zeiss optical microscope. Although the built-in illuminator and binocular eyepieces make it much more convenient to use than its older counterpart (see Figure 1.6), it is optically no better. The modern microscope focuses by moving the stage, whereas the older one moves the tube carrying objectives and the eyepiece. The arrow indicates the Bertrand lens, which permits the back focal plane (BFP) to be viewed through the eyepieces. (B) Image of a diatom (*Pleurosigma angulatum)* taken on this microscope. (C) The BFP seen using the Bertrand lens with the same sample in position. Since the sample has a regular pattern, it scatters light in particular directions, and this pattern is visible at the BFP where light is sorted by direction. (The condenser must be closed down to give parallel illumination to see the pattern.)

and testing the resolution of the microscope, and it also opens the door to a range of important contrast-enhancing techniques, which will be covered in Chapter 2.

GOOD RESOLUTION

How do we define how good a microscope, or a lens, is? Magnification does not really answer the question; it is no good forming a huge image if it is not sharp. (Microscopists often describe this as "empty magnification.") The important thing must be the ability to image and distinguish tiny objects in our specimen, since that is why we use a microscope in the first place. Two brilliant 19th century scientists turned their attention to the problem and came up with rather different treatments. This polarized the microscope community and arguments raged as to which was right. The protagonists themselves, John William Strutt (later Lord Rayleigh) and Ernst Abbe, looked on in bewilderment since it was clear enough to them that they were simply describing different imaging modes. Both approaches are correct, and both are relevant and important to the 21st century microscopist since they describe the two common ways in which we operate a microscope.

Resolution: Rayleigh's Approach

Rayleigh's approach to the resolution question (Rayleigh, 1880) was taken from astronomy and starts from a very simple situation: a lens forming a magnified image of an infinitely small point that emits light (Figure 1.10). What will the image of such a point source be?

The image is not an infinitely small point but a disk (Figure 1.11) called an *Airy disk* after Sir George Airy, the astronomer who first described it. The disk is surrounded by a series of progressively fainter haloes. The haloes only account for a small fraction of the light (the disk is 97%; first bright ring, 1.7%; second bright ring, 0.4%; third, 0.17%; fourth, 0.075%) so for most purposes they are ignored. But they can pop up and bite us, so never forget that they are there.

What causes this? The lens accepts only part of the wavefront leaving the point, so it is clear that we are not capturing all possible information about the specimen. What actually spreads out the light into a disc instead of a point is *diffraction*—scattering of light at the edge of the lens. The dark space between the disk and the first bright ring corresponds to the distance from the optical axis where rays from one side of the lens are out of step with rays from the other side. So the size of the disk depends on the size of the *lens*. To be precise, it depends on the angle subtended by the lens (i.e., the proportion of the total light that the lens can collect), so a small, close lens is equivalent to a larger one farther away. The radius of the disk, r, is given by

$$r = 0.61\lambda/\sin\theta$$

where θ is the half-angle of acceptance of the lens. (The half-angle is the angle between the optic axis and the edge of the lens, as shown in Figure 1.12.) We measure the angle at the object side and thus the formula gives the equivalent size of the disk at the specimen. In fact, since the ray paths work both ways, it also defines the smallest size of spot of light the lens could form at that plane, which is useful with scanning microscopes. We could use exactly the same formula to work out the size of the Airy disk at the image, using the angle at the other side of the lens, which will be smaller since the

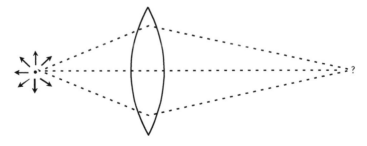

FIGURE 1.10   A point source of light is imaged by a lens. Will the image also be a point?

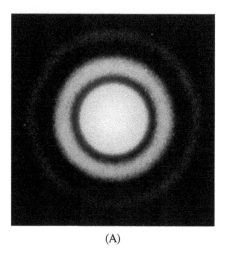

(A)

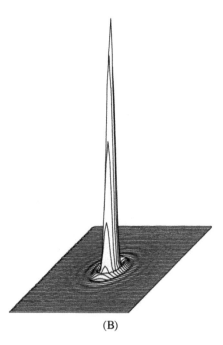

(B)

FIGURE 1.11 (A) An Airy disk. The limitations of printing make the central disk seem less bright than it is. (B) Graph showing the energy in central disk and rings. This gives a better impression of their relative intensities. (From Kriete, A., 1994, *Microscopy. Ullmann's Encyclopedia of Industrial Chemistry*, vol. B6, 5th ed., Weinheim, Germany: VCH Verlagsgesellschaft. With permission.)

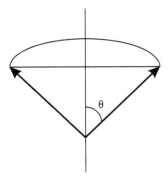

FIGURE 1.12   The half-angle of acceptance of a lens.

image is farther away. In microscopy, the useful property to us is the size of the Airy disk (or the size the Airy disk corresponds to) in the specimen, so that is what is used.

One surprising point is clear from this: there is no limit to the size of object we can see in a light microscope provided it is "a source of light," that is, light on a dark background. It will just appear as an Airy disk. However, if this tiny object is smaller than the Airy disk the image will tell us nothing about its size, shape, or structure.

Resolution is not a measure of how small an object we can see, it is the ability to distinguish two objects that are close together. So how can we define resolution?

Rayleigh's criterion for resolution was a rule of thumb devised by Rayleigh (1880) to provide a simple estimate for the case where two objects can just be resolved. He proposed that you could still distinguish two Airy disks where the center of one lies on the first minimum of the next (Figure 1.13), even though there is a considerable degree of overlap. Using this criterion, the equation for the radius of the disk can now be taken as specifying the resolution (*r* now standing for the *minimum resolved distance* as well as radius).

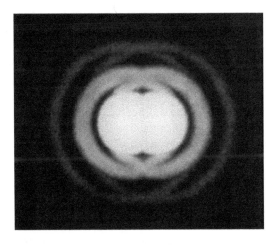

FIGURE 1.13   Two Airy disks with the edge of one on the center of the other (Rayleigh's criterion).

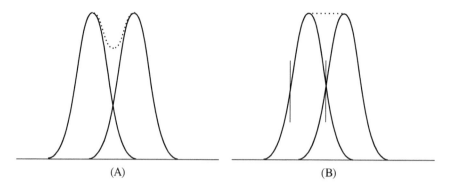

(A)   (B)

FIGURE 1.14   (A) Intensity plot through two Airy disks separated by Rayleigh's criterion. The dotted line indicates the combined intensity where they overlap. The intensity at the point of overlap is roughly three-fourths of the peak. (B) Two Airy disks separated by full width at half maximum (FWHM), indicated by the vertical lines. There is no dip in intensity between them.

What will the resolution be on this basis? Taking λ as 550 nm (center of the spectrum) sin θ as 0.65 (half angle 40.5°; acceptance angle 81°), a typical value for a student quality ×40 objective, we can work it out:

$$r = (550 \times 0.61)/0.65 = 516 \text{ nm}$$

In other words, a fairly ordinary, everyday lens will resolve objects down to the wavelength of the light used, which is a useful point to remember. We can do better than this, but more of that later.

Rayleigh's criterion is arbitrary. What it really defines is the *contrast* with which a particular level of resolution is transmitted. When two points of equal brightness are separated by the Rayleigh resolution, the intensity between the two points drops to 0.735 (roughly three quarters) the peak intensity (Figure 1.14A). Clearly this will be much easier to see if the two objects stand out well from their background than if they are dim. With a low contrast image we may not be able to resolve such fine detail; with a really contrasty one we might do better. Nevertheless, in practice Rayleigh's criterion accurately predicts the level of detail we can resolve.

An alternative criterion, used mainly because it is easy to measure, is full width at half maximum (FWHM). This is the width at half of the peak intensity (Figure 1.14B). It is a more optimistic criterion since there will be no dip in intensity between the two Airy disks but describes the limiting case beyond which we will not resolve two objects, whatever the contrast.

Rayleigh's approach treats the object as a collection of tiny luminous points. This is an accurate description of the situation in fluorescence microscopy; each molecule of a fluorochrome is very much smaller than the resolution of the microscope and thus effectively a point, and each emits light independently of every other molecule. In fluorescence microscopy Rayleigh's approach is the correct one to use.

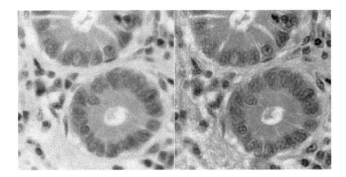

FIGURE 1.15   These two images were taken from the same sample without touching the focus or anything else in the imaging system. The only change was in the illumination, but it has made a dramatic difference to the quality and resolution of the final image.

Abbe

When we look at a stained section the situation is quite different. Every point is illuminated by light from the same source, so the points cannot be regarded as independent. It is easy to see (Figure 1.15) that in this case changing the illumination also makes a big difference to the resolution and the quality of the image.

This rather surprising observation caused considerable puzzlement among microscopists in the late 19th century. By this stage the major aberrations had been adequately corrected, and the development of microtomes had progressed to the point where sections able to exploit the full resolution of the microscope were routine. Yet microscopists often could not obtain the resolution their objectives should have been delivering. This problem was investigated in detail by the founder of modern light microscopy, Ernst Abbe, a young Austrian physicist who became a partner in the firm of Carl Zeiss in the late 19th century. This led him to propose his diffraction theory of image formation (Abbe, 1873).

When a light falls on a small object or on the edge of a larger one, it is diffracted, scattered in all directions. This was already common knowledge in Abbe's time. Figure 1.16 shows a row of tiny "point objects," spaced at equal distances apart and illuminated by parallel rays of light. Each particle will scatter light in all directions. However, if we look at the rays diffracted by each particle at any one angle, α, we notice something rather interesting. Each will be out of step with its neighbor by a distance, r. If r is half a wavelength ($\lambda/2$), then when all the rays at this particular angle are brought to the same point, destructive interference will take place; the point will be dark. If, on the other hand, the angle is such that r equals λ, one wavelength, or an integral number of wavelengths ($n\lambda$), then constructive interference will take place giving a bright spot. As we know, this is precisely what happens at the back focal plane of the lens. Thus we will end up with a series of bright and dark areas—a diffraction pattern—at the back focal plane, with each bright spot corresponding to a point at which $r = n\lambda$ (Figure 1.9).

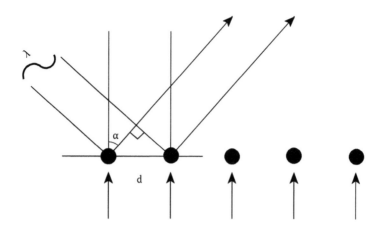

FIGURE 1.16   Diffraction of light at a row of point objects.

Our specimen is, in other words, a diffraction grating. It is simple enough to calculate the path difference, r, corresponding to a particular scattering angle, α. In the
diagram, the angle between the scattered ray and the specimen plane is $90 - \alpha$, so
that the angle between the specimen plane and the perpendicular between the rays
is again α. Thus, $\sin \alpha = r/d$, where d is the spacing between the object points, or
$r = d \times \sin \alpha$.

So, for a path difference (r) of λ, corresponding to the smallest diffraction angle
giving constructive interference,

$$\lambda = d \times \sin \alpha$$

It is clear that the smaller the spacing d, the larger the angle α must be to give a path
difference of one wavelength.

We do not actually need to have an array of points. Any two points will form diffracted beams at angles of $+\alpha$ and $-\alpha$. The closer together these two points are, the
larger will be the angle α. Abbe therefore treated the object as a series of pairs of
points, each with its own value of d and its own diffracted beams. The undiffracted
central beam, since it consists of light that has passed between the points, would not,
he argued, carry any information about the specimen. In other words, we will not
get any information about two points unless their diffracted beam can enter the lens.

So two points on our specimen, at distance d apart, will give a first-order dif-
fracted beam at an angle α, such that $\sin \alpha = \lambda/d$. If $\sin \alpha$ is small enough for the
beam to enter the lens, the two points will be resolved as separate; if not, they will
not be distinguishable from each other (Figure 1.17). The key parameter, therefore, is
the half-angle of acceptance of the lens, usually denoted by θ. If $\alpha > \theta$, the diffracted
light will not enter the lens; if $\alpha = \theta$, it will enter. We thus have a limiting value
of d—the minimum resolved distance—where $\alpha = \theta$, and $d = \lambda/\sin \theta$.

But, if we no longer illuminate the specimen with parallel light, but instead use a
converging beam of light, whose half-angle is equal to, or greater than θ, the situation is changed, as shown in Figure 1.18. Now light diffracted at any angle (α) up to

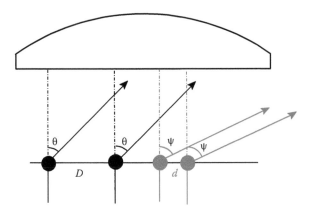

FIGURE 1.17   Diffracted rays up to angle θ will enter the lens, giving a minimum resolved distance of *D*. Particles closer together (*d*) diffract light at an angle ψ which will not enter the objective.

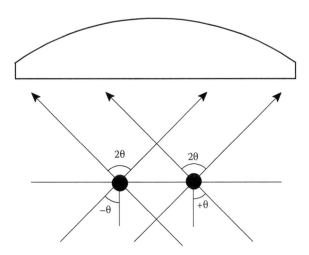

FIGURE 1.18   With convergent illumination rays at angles up to 2θ can enter the objective, so the resolution is twice as good.

α = 2θ can enter the objective. With such an illumination system, then, our minimum resolved distance is given by

$$d = \lambda/2\sin\theta$$

(Abbe's formula for the resolution of a microscope lens.)

With parallel light we will actually see the effects of diffraction at the back focal plane, as in Figure 1.9. This will no longer happen with converging illumination since there will be an infinite number of diffraction patterns, each corresponding to one incoming beam direction.

In spite of the difference between their approaches, both Abbe and Rayleigh came up with similar formulae and in both cases it is the half-angle, θ, which is all-important. The difference is that with transmitted light we will throw away half our resolution if we do not illuminate the specimen properly.

Visible light comprises a range of wavelengths (~400–700 nm), and although the color of light used will affect the final resolution, the range over the total visible spectrum is rather less than a factor of 2. There is, therefore, little prospect of improving resolution by using a shorter wavelength, especially since the eye is not very sensitive to violet and blue light between 400 and 500 nm. It is far more important to keep sin θ as large as possible. The absolute limit is where θ = 90° so that sin θ = 1. This is not very useful in practice, since the sample would be touching the lens. However, we can approach it—sin θ = 0.95 corresponds to θ = 72°, which is achievable and the 5% difference in resolution is insignificant. With λ = 550 this works out to a resolution, d (minimum resolved distance), of 290 nm.

Rayleigh's criterion is just a rule of thumb to estimate the case where the contrast becomes too low to distinguish two objects as separate, but Abbe's is absolute; if the corresponding diffracted rays cannot enter the objective, two objects cannot be resolved. Physicists describe Rayleigh's approach as incoherent imaging since each molecule of fluorochrome in a fluorescent sample is emitting light quite independently of its neighbors and out of step with them. Abbe's is called coherent or partially coherent. Even though we are not using a laser to illuminate the sample, the two adjacent points are receiving light from the same light source, so the waves hitting them at any instant are in step and can interfere.

ADD A DROP OF OIL...

So far we have talked about the wavelength of light as if it were immutable, but this is not strictly true. What is actually constant and immutable is the frequency of light (f), the rate at which it vibrates. It is the frequency, not the wavelength, that makes blue light blue and red light red. The wavelength is dependent on the frequency and the velocity with the simple relationship $v = f\lambda$ (obvious if you think of it as a series of waves going past a point). Some values of the frequency are $f = 4 \times 10^{14}$ (red); 5×10^{14} (yellow); and 7×10^{14} (blue).

The velocity of light is 3×10^{10} cm s^{-1} in vacuum, and about the same in air, but it travels more slowly in denser media. How much more slowly is given by the *refractive index* (n) of the medium, since n is the ratio of the speed of light in the medium to its velocity (v) in vacuum or air.

$$n = v \text{ in vacuum}/v \text{ in medium}$$

It follows that in glass (n ~1.5) the velocity of light is considerably lower and its wavelength is correspondingly shorter. Figure 1.19 shows how the slowing down of the light brings the waves closer together. If we can image our sample in a medium of this refractive index we should be able to get better resolution. Glass is obviously not convenient, but many mounting media of approximately the correct index are available and with these we can indeed get the expected resolution improvement.

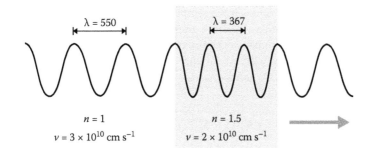

$\lambda = 550$

$\lambda = 367$

$n = 1$
$v = 3 \times 10^{10}$ cm s^{-1}

$n = 1.5$
$v = 2 \times 10^{10}$ cm s^{-1}

FIGURE 1.19 When light enters a dense medium it slows down, and since the frequency of vibration stays the same, the wavelength becomes shorter.

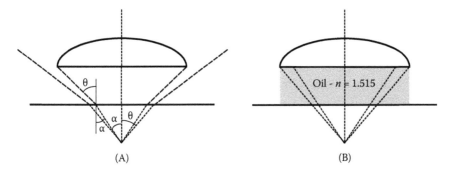

Oil - $n = 1.515$

(A)   (B)

FIGURE 1.20 When a specimen is in a high refractive index medium, the lens will not receive light from the whole angle θ but only from the smaller angle α, because of refraction at the top of the coverslip. The relationship θ/α is the refractive index (Snell's law). With oil between the lens and coverslip rays throughout the whole angle θ can enter the lens.

There is one caveat in this: the high refractive index medium must go all the way to the lens. Otherwise, as Figure 1.20A shows, there is a reduction in the acceptance angle from θ to the smaller angle α due to refraction at the glass–air interface. This also depends directly on the refractive index (Snell's law of refraction says that sin θ/sin α = n), and therefore the reduction in angle will exactly cancel out the gain from the reduction in wavelength. To avoid this, as shown in Figure 1.20B, we use immersion oil between the lens and the coverslip. Immersion oil is chosen to have the same refractive index as glass; the standard value is 1.515.

Although the Rayleigh and Abbe equations are still literally true, since it is the wavelength that is reduced, we are accustomed to using the conventional (vacuum) wavelength. Therefore, to cope with different media, we add the refractive index in which the lens operates into the formula. The Rayleigh formula then becomes $r = 0.61\lambda/(n \sin \theta)$. The composite $n \sin \theta$ is called the numerical aperture (NA) of the lens and is always written prominently on every objective. Similarly, the Abbe formula becomes $r = \lambda/2NA$. (*Numerical* aperture because it is a dimensionless number, a ratio, and therefore has no units.)

Since sin θ must always be less than 1, the NA is always less than the refractive index the lens works with (normally 1.515 for oil). Likewise, if we see a numerical aperture greater than 1 we know it is an immersion objective, designed to work in oil, water, or glycerol. In practice a research grade oil lens will have an NA of 1.4 (sin θ = 0.92, or θ = 67.5°), and "student" oil immersion lenses typically offer NA 1.3, which is not much worse. Lenses with even higher NA are made for specialist applications such as total internal reflection fluorescence microscopy (TIRF) (Chapter 16). What resolution will this give us?

$$r = (0.5 \times 550)/1.4 = 196 \text{ nm}$$

less than half the wavelength of the illuminating light.

Often in cell biology we want to look at living cells in water. The refractive index of the medium in this case is 1.3, so our gain in resolution will be less, but a water-immersion lens will still give us better resolution than a dry lens. An equivalent water-immersion lens (acceptance angle = 67.5°) will give us an NA of 1.2, and these are popular (though expensive) for work on living cells. A water-immersion lens can be designed to work either with or without a coverslip; a parallel piece of glass does not affect the angle of the rays entering and leaving it if the refractive index is the same on either side. But as we will see in Chapter 7, the lens can only be designed for one case or the other.

KÖHLER ILLUMINATION

There are several possible ways of setting up illumination in a microscope to fulfill the requirements for Abbe imaging, but the one universally used in research-quality microscopes is Köhler (Koehler) illumination (Figure 1.21). This was invented by August Köhler (1893, 1894), who was invited to join the Zeiss company on the strength of this work. In Köhler illumination a lens—the lamp lens or lamp con-denser—is placed in front of the lamp. The back focal plane of this lens is used as the source of light to illuminate the specimen. This guarantees even illumination since whatever the lamp is like, its light will be evenly distributed in the back focal plane. At this plane is an iris diaphragm—the field iris—which controls the size of the effective source and therefore the size of the illuminated area on the specimen.

The field iris, and hence the back focal plane of the lamp lens, is focused on to the specimen by the substage condenser. At the back focal plane of the substage condenser is another iris diaphragm that will (as should now be obvious) control the angle of the light reaching the specimen and must therefore be adjusted for different numerical aperture objectives.

This system, while not the cheapest option thanks to the number of lenses involved, gives the microscopist total control of the important factors in the illumi-nation. The field iris controls the area illuminated; the condenser iris the angle of illumination. Its main disadvantage is that at high magnifications, just when we need most light, we are using only a small part of the available illuminated area. With modern light sources this is really not an issue, though it must have been a problem in the days of Abbe and Köhler.

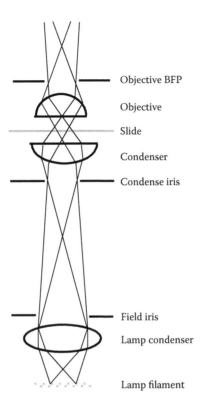

Objective BFP

Objective

Slide

Condenser

Condense iris

Field iris

Lamp condenser

Lamp filament

FIGURE 1.21   Köhler illumination. The paths of two rays of light from the lamp filament are shown; they form an image of the filament at the condenser iris and then again at the objective back focal plane. At the field iris the rays are spread over the back focal plane and rays at the same angle cross at the same point; the same happens at the specimen.

REFERENCES

Abbe, E. 1873. Beiträge zur Theorie des Mikroskops und der mikroskopischen Wahrnehmung. *Archiv für mikroskopische Antomie*, 9, 413–420.

Köhler, A. 1893. Gedanken zu einem neuen Beleuchtungsverfahren für mikrophotographische Zwecke. *Zeitschrift für wissenschaftliche Mikroskopie*, 10, 433–440.

Koehler, A. 1894. New method of illumination for phomicrographical purposes. *Journal of the Royal Microscopical Society*, 14, 261–262. (Condensed translation of Köhler, 1893)

Kriete, A. 1994. *Microscopy. Ullmann's Encyclopedia of Industrial Chemistry*, vol. B6, 5th ed. Weinheim, Germany: VCH Verlagsgesellschaft, pp. 213–228.

Rayleigh, Lord. 1880. Investigations in optics, with special reference to the spectroscope. 1, Resolving, or separating, power of optical instruments. *Philosophical Magazine* (Series 5), 8, 261–274.

2 Optical Contrasting Techniques

Staining a specimen is not always convenient or even possible, especially in the case of living cells. It is clear that all the structures inside the living cell differ in some way from their surroundings. Why, then, are we often unable to see them in the microscope? The answer is that they differ from their surroundings in refractive index but not in the amount they absorb light. To make these structures visible, we need some way to convert these differences in refractive index into differences in amplitude (color or intensity) in the image.

DARKFIELD

A structure that differs in refractive index from its surroundings still scatters (diffracts) light, even if it does not absorb it. If we form an image using only scattered light, such structures will be visible, even if they scatter only weakly. They will be bright, whereas the background, which scatters no light, will be dark—hence the name *darkfield*.

Remembering that parallel rays of light all pass through the center of the back focal plane (BFP; Chapter 1), we could use parallel illumination and put a piece of clear glass with a central black spot at the BFP of the objective. This would work from the point of view of contrast. However, as Abbe demonstrated, with parallel illumination (Chapter 1) we are not getting the maximum resolution from our microscope.

A more effective scheme is to use a hollow cone of illumination (Figure 2.1). We replace the condenser diaphragm with a ring-shaped aperture that is so large that the light from it falls outside the acceptance angle θ of the objective lens. Light scattered up to an angle of 2θ can enter the objective, so we do not lose any resolution. It follows that we must use a condenser with a larger numerical aperture than the objective. No resolution is lost if this criterion is met, but it means that we cannot use darkfield with the highest-power objectives. We may even have to use an oil-immersion condenser with a high-power dry objective. Some oil-immersion lenses are fitted with iris diaphragms so that their numerical aperture (NA) can be reduced for darkfield imaging. This wastes performance, but it makes it easy to switch between, for example, fluorescent imaging at full resolution and darkfield at lower resolution. Because the light from the back focal plane is evenly distributed at the image plane, the field of view is fully and evenly illuminated as normal (Figure 2.2), provided that the condenser is correctly focused, which requires care.

The simple condenser and ring-stop shown in Figure 2.1 waste a lot of light, so the image is dim. Various condensers have been developed that concentrate the light into the periphery of the BFP rather than just blocking it, thus giving a much

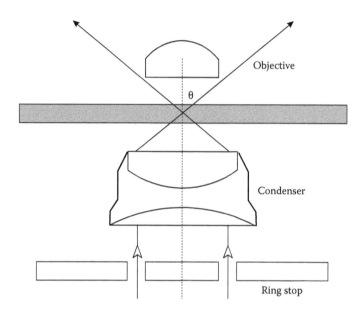

FIGURE 2.1 A simple darkfield illumination system. The ring-stop at the condenser BFP allows light with a narrow range of angles all larger than the acceptance angle θ of the objective to illuminate the specimen. The field of view is fully illuminated, but no direct light enters the objective.

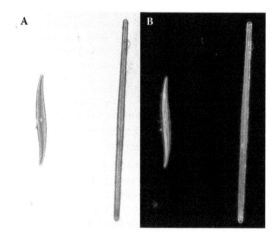

FIGURE 2.2 Diatom frustules: (A) in brightfield and (B) in darkfield, taken with ×10 objective.

brighter image. For low-magnification darkfield, fiber-optic ring illuminators can be very successful. The snag with using a special condenser is that it is inconvenient to switch back to regular illumination. Another alternative is incident-light (epi-) darkfield (Figure 2.3), in which the illuminating light is transmitted through a sleeve built into the exterior of the objective and focused by ring mirrors. This avoids the

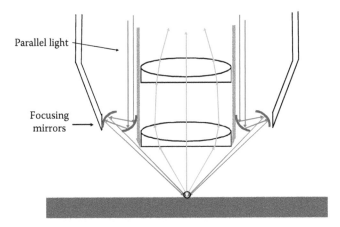

Parallel light

Focusing
mirrors

FIGURE 2.3 Diagrammatic layout of an epidarkfield objective. Parallel light is sent down the periphery of the objective and tightly focused by a pair of ring mirrors. This is effective for both opaque and transparent specimens.

condenser problem and also allows imaging of opaque samples; very tiny structures on surfaces can be detected in this way.

For many years, darkfield was the only technique available for introducing contrast into unstained specimens. Apart from its resolution limitation, darkfield is vulnerable to stray dust or other out-of-focus objects, which scatter light into the objective, destroying the contrast. Scrupulous cleanliness is essential. However, because it allows you to obtain a huge amount of contrast enhancement, darkfield is still a highly useful technique for objects that are otherwise hard to see.

PHASE CONTRAST

In the 1930s, the Dutch physicist Zernike (1942a, 1942b) hit upon the idea of introducing a phase shift between scattered and undiffracted light so that they could interfere with each other. If we make the diffracted rays just half a wavelength out of step with the direct rays, anything in the specimen that scatters light appears dark in the final image. We do not need to stop the direct rays from entering the lens, and we end up with a more natural-looking image because our structures appear dark on a light background, as in a stained brightfield sample. In principle, there should be no limitation on resolution.

The physics of scattering tell us that the phase of the scattered light is retarded by ¼λ, so we need to retard it by a further ¼λ to bring the path difference to ½λ. In practice, this will be approximate because the different refractive indices of the cell components also affect the final phase, and this is responsible for the variations of contrast in the final image.

To separate the diffracted and direct rays, we use an annular stop in the condenser, just as for darkfield, except that now we pick a size such that all the light does enter the objective lens. Now we know that all the direct light will be in a ring

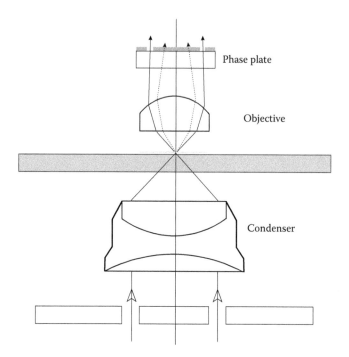

Phase plate

Objective

Condenser

FIGURE 2.4 The optical layout of Zernike phase contrast. Direct light enters through a ring-stop and, as in darkfield, illuminates the whole of the field of view. At the objective BFP, direct light will once again be in a ring, whereas scattered light (dashed) will be everywhere. A phase plate at the BFP can therefore introduce a phase shift between direct light and scattered light.

at the back focal plane of the objective. The diffracted rays, on the other hand, will be everywhere (Figure 2.4).

At the objective BFP, we place a piece of glass, with a carefully made groove in it matching the ring where the direct light passes (Figure 2.5). Light travels more slowly through glass than through air (Chapter 1). Most of the diffracted light passes through the full thickness of the glass, so diffracted light is retarded relative to the direct rays that have gone through the thinner part.

How deep must the groove be? If the refractive index of glass is 1.5 and that of air is taken as 1, the path difference will be the depth, t, times the difference in refractive indices (1.5 − 1, or 0.5), so the path difference is $0.5 \times t$. Taking a wavelength of 550 nm (the center of the spectrum), we want the path difference to be $\frac{1}{4}\lambda$, 550/4, which is 137.5 nm. The depth, t, will therefore be twice this number: 275 nm. Clearly, such accuracy requires some precision to achieve, and phase contrast objectives were once very expensive, although modern production techniques (evaporating a layer rather than polishing a groove) have now made the cost no higher than normal objectives.

The direct light is often brighter than the diffracted light, so usually an absorbing layer is added at the base of the groove to bring them to a more equal intensity. This is somewhat arbitrary, because different specimens will scatter more or less light, so

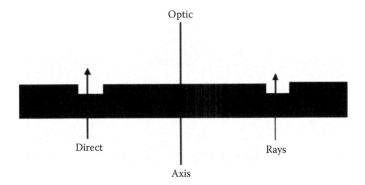

FIGURE 2.5   The phase plate at the objective BFP has a groove in it so that the direct light passes through less glass than the scattered light.

different degrees of absorbance are made. Phase lenses are usually marked as having high or low absorbance.

Because our groove can have only one depth, we will only have ¼λ path difference for one color (550 nm); therefore phase contrast works best with a green filter. A green filter is not essential—after all, we started out with only an estimate of the path difference introduced by the specimen—but generally we will get the best contrast using the green filter.

The phase ring must be a substantial fraction of the diameter of the BFP if we are not to sacrifice resolution through a too-small illumination angle. This can sometimes be compromised when one ring-stop does duty for several different objectives, but otherwise phase contrast is a high-resolution technique. So long as the path differences within the sample are small the final contrast will depend directly on the refractive index (Figure 2.6) making the image "normal" looking and easy to interpret (Figure 2.7A). However, if samples are very thick or highly refractile, the path difference can go past ½λ to 1λ or 1½λ. This causes contrast reversals, as can be seen in Figure 2.7B. This means that phase contrast is best for relatively thin samples, such as cell cultures.

POLARIZATION

"Normal" white light from a lamp vibrates in all directions. However, certain very anisotropic materials can filter out all but one plane of vibration (Figure 2.8A), giving *plane-polarized* light. These days the material used is synthetic, but polarization happens to some extent whenever light is reflected from a nonmetallic surface (which is why sunglasses are often polarized). A second polarizer, at right angles to the first, blocks the light completely. If the second polarizer is rotated from the blocking (*extinction*) position, it progressively transmits more and more light; the light always follows the orientation of the second polarizer.

Many crystalline and fibrous materials are *birefringent*—that is, their refractive index is different for two directions of polarization. If we pass plane-polarized light through a birefringent material, the light splits into two beams, polarized at right

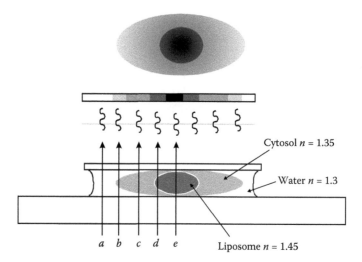

FIGURE 2.6 How refractive index determines the final contrast in a phase-contrast image. Ray *a* just passes through water, its phase is shifted a bit, and this gives the background intensity for the image. Ray *b* passes through the edge of the cytosol, so its phase is retarded a bit more, and the image appears darker. As rays pass through more of the cytosol they are progresssively retarded more (e.g., ray *c*) so the thicker the cytosol the darker the image. When a ray reaches the highly refractile liposome (*d*) its phase is shifted further and we see an abrupt change in contrast. The center of the liposome (ray *e*) will be the darkest part of the image. The contrast on the right-hand side exactly mirrors that on the left.

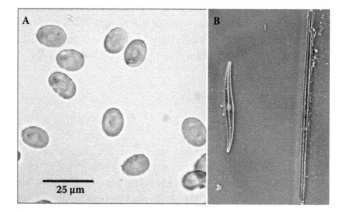

FIGURE 2.7 (A) Erythrocytes (red blood cells) of turtle under phase contrast. These are ideal subjects for phase and the contrast is formed very much as shown in Figure 2.6. (Reproduced from G. Benga et al., in press, *Bulletin of Molecular Medicine*. With permission.) (B) The same diatom frustules shown in Figure 2.2 imaged in phase contrast using a green filter. Because these are both thick and highly refractile, we get very large phase shifts so that now the thickest parts appear light rather than dark.

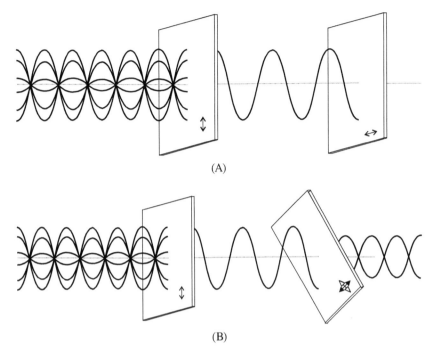

(A)

(B)

FIGURE 2.8 (A) Light that is not polarized vibrates in all directions. After passing through a polarizer, the light is constrained to vibrate in one direction only. A second polarizer with its axis at right angles to the first blocks the light completely. (B) When plane-polarized light passes through a birefringent substance, it is resolved into two components whose axes are determined by the crystal orientation and whose magnitudes depend on the angle they make with the original polarization.

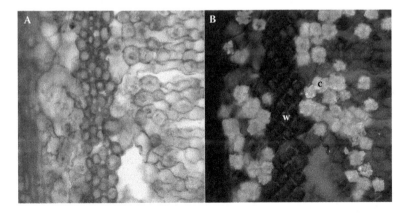

FIGURE 2.9 Brightfield and polarized light micrograph of the stem of *Tilia* (lime, linden). Crystalline inclusions (c) in the parenchyma cells appear very bright between crossed polarizers, and the birefringent cellulose in the thickened cell walls (w) also shows up very strongly.

angles to each other (Figure 2.6B). The angle of polarization is determined by the optic axis of the birefringent material, and the strength of each ray is determined by the angle it makes with the original plane of polarization. Thus, if the birefringent material is at 45° to the original polarization, we get two equal output beams, both at 45° to the original direction. As we will see later in the following section on *differential interference contrast* and in Chapter 5, this property lets us perform all sorts of useful optical tricks. But we can also use it as a contrasting mechanism, making birefringent materials in cells visible. With a polarizer placed in front of the lamp and another at right angles before the eyepiece, all light should be blocked. But if there is any birefringent material in the sample, the plane of polarization is changed and light *will* pass through the second polarizer (often called the *analyzer*). Collagen shows up particularly bright in polarized light, as does cellulose in plant cells (Figure 2.9). Many other components are birefringent, and at high magnifications we can see microtubules and actin bundles.

Sixty years ago, polarized light microscopy was a key technology in cell biology, responsible for much of our knowledge of the cytoskeleton and other structural proteins. Now, biologists use other techniques, and polarization microscopy is mainly used by mineralogists, for whom it is an important quantitative tool. However, advanced polarization techniques for cell biologists are again on the market (Oldenbourg, 2004), and this technique may return to the mainstream.

DIFFERENTIAL INTERFERENCE CONTRAST

The *differential interference contrast* (DIC) technique was introduced in the 1950s by the Polish–French scientist Nomarski (Normarski & Weill, 1955) and is often known by his name. Its basic principle is simple: to make each ray of light interfere with another passing through the specimen a very small distance away. If the refractive index of the specimen is changing at that point, there will be a path difference between the two rays, whereas if the refractive index is uniform there will not be a path difference. The contrast we see in the final image depends on the local *rate of change* in refractive index of the specimen, hence the name *differential interference contrast*. The idea is to keep the two rays' separation less than the resolution of the microscope, so that resolution is not compromised; it should be as high as any other widefield imaging technique.

The practical arrangement of DIC, illustrated in Figure 2.10, is more complex than phase contrast but is not difficult to understand. Below the condenser is a polarizer, so that plane polarized light enters the system. At the condenser's back focal plane is a Wollaston prism: two quartz wedges, with their crystal axes at right angles, cemented together to form a block. Each axis is set to be at 45° to the polarization of the incoming light. Because quartz is birefringent—it has different refractive indices along and across the crystal axis—it resolves each incoming ray of polarized light into two rays of mutually opposite polarization, emerging at slightly different angles.

Because a difference in *angle* at the BFP translates to a difference in *position* at the image plane, the condenser lens focuses these two rays to slightly different points on the specimen. These points must be closer together than the resolution of the microscope, so that we do not see a double image.

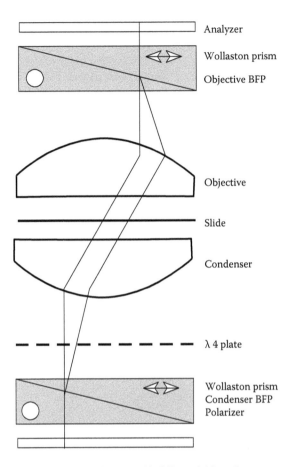

Analyzer

Wollaston prism

Objective BFP

Objective

Slide

Condenser

λ 4 plate

Wollaston prism
Condenser BFP
Polarizer

FIGURE 2.10   The optical layout of Nomarski differential interference contrast.

At the back focal plane of the objective is another, identical Wollaston prism. This prism recombines the rays so that they travel together again. It also reverses the path difference between the two rays introduced by the first prism. (Some rays will have been through a thick part of one orientation and a thin bit of the other, others the opposite.) Because rays going through the left side of the condenser prism pass through the right side of the objective one, this arrangement exactly corrects these path differences, so we are left only with the path differences caused by the specimen (which are what we are interested in).

These path differences would not be enough in themselves to create interference, so we introduce a further path difference between the two rays using a flat sheet of a birefringent substance often called a $\lambda/4$ plate, since it produces a quarter wave shift between the two beams. There is usually some additional mechanism for manually varying the path difference between the two beams, so that we can adjust it to give optimal contrast with our particular specimen. This adjustment may be made by rotating the polarizer or by displacing one Wollaston prism relative to the other.

Although our two rays have now been brought back together, they cannot yet interfere with each other because they are still polarized perpendicularly to each other. Above the objective there is, therefore, another piece of Polaroid, the analyzer, set at 45° to the two rays (and 90° to the initial polarizer). This brings both rays back to the same polarization, so that interference can take place in the final image.

The image produced by differential interference contrast has a striking bas-relief appearance, with one side of each object appearing bright and the other shaded. This effect results from the directionality of the technique: The refractive index *increases* on one side of the object and *decreases* on the other (Figure 2.11). It is important to realize that this is just a contrasting technique and does not reveal actual three-dimensional information about the sample. The path-length differences introduced in DIC do not usually exceed one wavelength, so the contrast reversals and haloes of phase contrast are absent. With thin samples, DIC may give too little contrast to be useful, but with thicker specimens it is often the method of choice. Most DIC implementations also offer the option of introducing a λ or "rose" plate into the optical path. Instead of a ¼λ path difference between the two polarized rays, there is now a much larger difference. This means that the path difference is a whole number of wavelengths for some colors and a fractional number for others, so the background

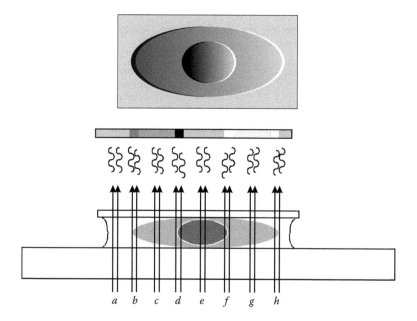

FIGURE 2.11 The contrast in a DIC image, using the same stylized cell as in Figure 2.6. The pair of rays *a* both pass through water and so stay in phase. When one ray goes through the cytosol and the other is in water (*b*) we get a large phase difference and strong dark contrast. The rays *c* go through slightly different thicknesses of cytosol, so still produce some contrast, and the boundary of the liposome will produce strong contrast. At the center of the liposome both rays stay in phase so no contrast is generated. One the right side of the image (rays *f*, *g*, and *h*) the phase differences are the same as in *b*, *c*, and *d*, but in the opposite direction—now the left ray is retarded. Contrasted parts therefore appear bright.

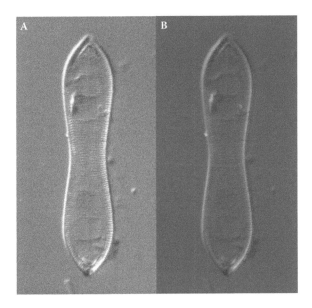

FIGURE 2.12 DIC images of a diatom frustule, ×40 objective. (A) Conventional DIC: Note the striking bas-relief appearance of the image. (B) With λ-plate introduced to give color contrast.

is colored. Then the path differences introduced by the specimen are also translated into color differences (Figure 2.12). Whether the color gives any additional scientific information is debatable, but it can be invaluable for creating striking poster displays and for entering micrograph competitions.

HOFFMAN MODULATION CONTRAST

Hoffman contrast (Hoffman & Gross, 1975), like DIC, images gradients in optical thickness (changes in thickness or refractive index) within the specimen, but unlike DIC it does not use the resulting change of phase to generate contrast. Instead, it makes use of the fact that a region where thickness or refractive index (RI) is changing will act like a little prism or lens and change the direction in which the light travels (Figure 2.13). This means it will change the position at which it appears in the back focal plane. Going from left to right, if the optical thickness is increasing the light will be deflected to one side of the BFP; if it is decreasing it will be deflected to the other. For this to be any use we must illuminate with parallel light, otherwise the result would be scrambled at the BFP (Chapter 1), so a slit is used at the condenser BFP. At the BFP of the objective is a plate that is dark on the left, gray in the middle, and clear on the right (Figure 2.14A). Light that has gone straight through will fall on the gray part, so it will continue with reduced intensity (usually 15%). Light that was deflected to the right will end up on the left of the BFP and its intensity will be reduced further. Light that was deflected to the left will end up on the right of the BFP and will be transmitted at full strength. The resulting image (Figure 2.15) will have a very similar appearance to a DIC image but the contrast has been produced in quite a different way.

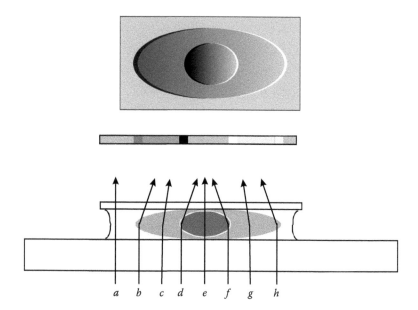

FIGURE 2.13   The principle of Hoffman Modulation Contrast, again using the stylized cell of Figure 2.6. Changes in refractive index in the sample will act as little lenses, deflecting rays that pass through. With parallel illumination, this means that they will appear at different places in the BFP. Ray *a* sees no change in RI, and so will not be deflected. Ray *b* sees a change of RI as it hits the edge of the cytoplasm, and is deflected. Passing through the cytosol, ray *c* sees a smaller change, and is deflected less. Ray *d* hits the edge of the liposome and is deflected a lot, while ray *e* is not deflected at all, like the ray that passes through the center of a lens. Rays *f*, *g*, and *h* show matching deflections, but in the opposite direction. The overall effect is very similar to DIC.

The simple system described thus far, which was the original form of Hoffman contrast, has the disadvantage that the parallel illumination halves the resolution of the system (Chapter 1). To get around this modern systems use a modified arrangement in which the slit is offset, so that the illumination, though still parallel, is very oblique. That means that rays diffracted at almost 2θ can still enter the objective, and effectively no resolution is lost. The modulation plate now has the gray stripe well over on the left (Figure 2.14B), and the dark zone is very small. This does not matter since very little light is transmitted through this zone anyway, and it means that the plate will interfere much less with other imaging modes.

The figure also shows an added refinement. Half of the slit has a strip of polarizer over it. Another, full-size polarizer just beneath can be rotated. If both polarizers are parallel the slit is full width, if they are crossed the strip covered by the polarizer transmits no light, so the slit becomes narrower. This gives us control over the amount of contrast introduced.

Even though the light going through the sample is now polarized, we get no polarization contrast since there is no analyzer. This gives Hoffman contrast an advantage when looking at birefringent material, such as tissue containing collagen or cellulose, which would show confusing polarization effects in DIC. This feature is

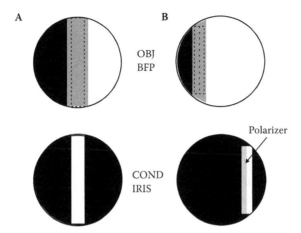

FIGURE 2.14 The optical arrangement for Hoffman contrast. (A) The original 1975 version. The specimen is illuminated by a slit at the plane where the condenser iris would normally be. This is imaged on the gray part of the mask at the objective BFP (dotted rectangle) so direct rays are darkened. Deflected rays that hit the left-hand part of the mask are effectively blocked, while rays deflected in the opposite direction pass through the clear area. (B) Since the original arrangement loses resolution, this more modern arrangement moves the slit to one side; the illumination is still parallel but oblique. Since less of the objective BFP is darkened, the final image is brighter but still has the same contrast. As a further refinement, a strip of polarizer covers half the slit. By putting a cross-polarizer in the light path before the condenser, the slit can be effectively narrowed, increasing contrast.

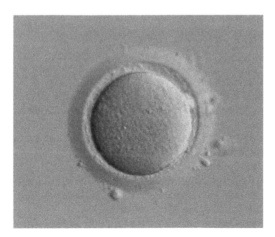

FIGURE 2.15 "Stripped" human oocyte; granulosa cells that had surrounded the oocyte have been removed. Hoffman modulation contrast. (Courtesy of RWJMS IVF Laboratory.)

particularly handy when looking at cells grown on plastic coverslips or in plastic Petri dishes, since these plastics are so birefringent that DIC imaging is impossible.

Hoffman imaging needs a turret with different slits for different objectives, and this will also have a blank space for widefield imaging. The technique works best in monochromatic light, since different wavelengths will be deflected by different amounts (*dispersion*; Chapter 7).

WHICH TECHNIQUE IS BEST?

Darkfield has two serious limitations: (1) it cannot use lenses of the highest numerical aperture, and (2) any dirt or dust in the system or objects outside the plane of focus scatter light into the background and degrade the image. Darkfield does, however, have a virtually limitless ability to enhance contrast; with a properly set up system, all direct light is excluded. No other contrast technique can match darkfield in this respect.

Phase contrast is best, as one might expect, when the specimen produces a phase difference of about ¼λ. Very thick or highly refractile objects produce strange and confusing reversals of contrast. Phase contrast is ideal with fairly thin and uniform specimens, such as monolayers of cells—nothing else can compare with it for these—and it also works well with unstained sections.

Differential interference contrast often does not give much contrast with the thin objects that are ideal for phase contrast; there just is not a great enough path difference between adjacent points. On the other hand, DIC is not bothered at all by thick and highly refractile specimens, because the overall path difference between one area and another means nothing—only the rate of change produces contrast. DIC is therefore the technique of choice for looking at protozoa, small organisms, and so on. DIC also can be applied in incident light microscopy, where it picks up, for example, height variations in a reflective surface or variations in thickness in a transparent film.

Hoffman contrast has mainly found favor for imaging cells on plastic substrates, where DIC will not work since the substrate is birefringent. For this it is ideal; without the "plastic problem" DIC will usually do better.

REFERENCES

Benga, G., Chapman, B.E., Romeo, A., Cox, G.C., and Kuchel, P.W. In press. Comparative NMR studies of diffusional water permeability of red blood cells from different species: XVII Green Sea Turtle (*Chelonia mydas*). *Bulletin of Molecular Medicine* (in press).

Hoffman, R., & Gross, L. 1975. Modulation contrast microscope. *Applied Optics*, 14, 1169–1176.

Nomarski, G., & Weill, A.R., 1955. Application à la métallographie des méthodes interférentielles à deux ondes polarisées. *Revista De Metalurgia*, 2, 121–128.

Oldenbourg, R. 2004. Polarization microscopy with the LC-PolScope. In *Live Cell Imaging: A Laboratory Manual*, D.L. Spector and R.D. Goldman, eds. Cold Spring Harbor, NY: Cold Spring Harbor Laboratory Press, pp. 205–237.

Zernike, F. 1942a. Phase-contrast, a new method for microscopic observation of transparent objects, part I. *Physica*, 9, 686–698.

Zernike, F. 1942b. Phase-contrast, a new method for microscopic observation of transparent objects, part II. *Physica*, 9, 974–986.

3 Fluorescence and Fluorescence Microscopy

Fluorescence microscopy is the definitive technique of cell biology. The histologist uses conventional transmitted light microscopy, and the anatomist uses electron microscopy, but the cell biologist uses fluorescence. The ability to label individual structures, molecules, or cell compartments gives us enormous power to visualize the structure and even the dynamics of the workings of cells. Immunolabeling identifies components with exquisite sensitivity in fixed (and therefore dead) tissue, whereas tagging with green fluorescent protein (GFP) and other expressible fluorescent proteins (Chapter 12) can reveal molecular information about living cells. Fluorescent in-situ hybridization (FISH) reveals genes. Add to this the dyes such as DAPI (DNA) and rhodamine 123 (endoplasmic reticulum or ER), which target specific cell structures and compartments; indicator dyes that can reveal ion concentrations and membrane potential; and the natural fluorescence of many important cell components (NADP, chlorophyll); and the fundamental power and importance of fluorescence microscopy are clear. Fluorescence microscopy also provides the underpinnings of more recent techniques, such as confocal and multiphoton microscopy, which now enable us to image in three dimensions, not just two.

To use fluorescence microscopy effectively, it helps to have a rudimentary understanding of light, optics, and the process of fluorescence, as well as how conventional and confocal microscopes work. Such knowledge puts you in a better position to choose appropriate dyes, select correct filters, figure out what is going wrong, and decide on the best technique to apply to your problem.

WHAT IS FLUORESCENCE?

Light is electromagnetic radiation: an arbitrarily defined part of the electromagnetic spectrum usually regarded as covering the wavelengths from around 200 nm to 2 μm. The visible spectrum lies between 400 nm and 700 nm; wavelengths of 700 nm to 2 μm are known as *infrared*, and wavelengths between 200 nm and 400 nm are called *ultraviolet* (UV). Even longer wavelengths are known as *microwaves*, and *radio waves* are longer still (up to 2 km). Shorter wavelengths are x-rays and gamma rays. The only difference among these forms of radiation is their energy: the higher the energy, the shorter the wavelength. So an x-ray photon (the quantum of electromagnetic radiation) carries a lot more energy than a visible light photon, and a microwave photon carries much less. Within the realm of light, it is important to keep in mind that UV and blue light are more energetic than red and infrared (Figure 3.1), because energy is fundamental to the process of fluorescence.

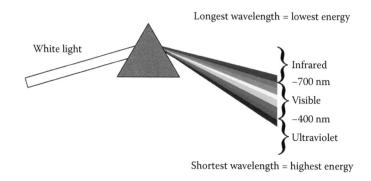

FIGURE 3.1   The visible spectrum.

The process that leads to fluorescence starts when a photon is absorbed by a suitable molecule, giving up its energy to an electron and thus boosting the electron to a higher energy state. In this context, "suitable" means that there is a possible electronic transition requiring an amount of energy close to that carried by the photon. (The fact that only discrete states can exist was discovered by Einstein in 1905.) The important thing to realize is that this interaction will happen only if the electron has a pretty low energy: outer shell-bonding electrons on long, conjugated organic molecules. Inner-shell electrons need much more energy to move—in the x-ray range—and x-ray fluorescence is also an important technique, giving us information about atoms rather than molecules, but x-ray fluorescence is beyond the scope of this chapter.

After a molecule absorbs light, it can return to the ground state via a number of processes. To explore these, we use a generic energy-level diagram, called a *Jablonski diagram* (Figure 3.2). A Jablonski diagram is a "cartoon" of the energy

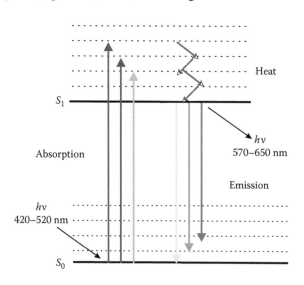

FIGURE 3.2   Jablonski diagram for absorption and fluorescence.

levels in a molecule, showing only the details that are important to describe the process of interest. Bold lines represent the fundamental electronic states, and lighter lines represent vibrational states (which can be thought of as the "temperature" of the electrons).

A vertical arrow represents absorption of light. Typically, the absorbed photon has more energy than the minimum to lift the electron to an excited (S_1) state, so the electron is raised to a vibrational level above that level. Before it returns to the ground state, the electron must reach the lowest excited state (Kasha's law). To do this, it must lose energy as heat, indicated by the zigzag lines in Figure 3.2. Once it is at the lowest vibrational level of the S_1 state, the electron can drop back to the ground state, which means it must give up a chunk of energy, which is emitted as a photon (fluorescence). Because some of its excitation energy has been lost by this stage, fluorescence always occurs at longer wavelength than absorption (Stokes' law). Figure 3.3 shows the range of absorbed and emitted wavelengths—the excitation and emission spectrum—for a typical fluorescent dye, in this case rhodamine.

Molecules have more than one excited electronic state, however. The electronic states of most organic molecules can be divided into singlet states and triplet states. In a *singlet state,* all electrons in the molecule are spin-paired, whereas in a *triplet state* one set of electron spins is unpaired. Figure 3.2 shows the simplest state, where an electron is promoted to the lowest excited state: S (singlet) 1. A higher-energy photon might elevate the electron to S_2 or even S_3 (Figure 3.4); the absorption spectrum will show peaks at each of these levels. The fluorescence spectrum, however, remains the same, because fluorescence occurs only from the lowest excited state. All higher states relax (by losing heat) to this state before fluorescing. The process of fluorescence takes time—a short, but measurable time known as the *fluorescence lifetime*—and we can measure and make use of this (Chapter 15).

Fluorescence is not the only way out for an excited electron. It may lose all its energy as heat (internal conversion), and no fluorescence occurs. The likelihood of this depends on the number of available vibrational states; obviously for a "good" fluorochrome we want internal conversion to be an unlikely process.

Another possibility is that the electron's spin may flip, so that the electron enters a triplet state. This is particularly likely to happen if the electron has been promoted to

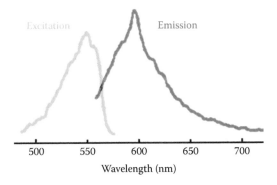

FIGURE 3.3   Absorption and fluorescence spectra of a dye molecule (rhodamine).

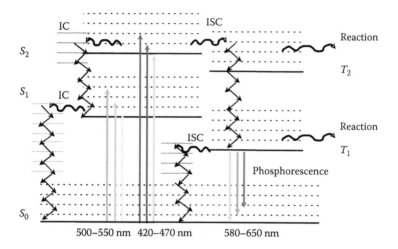

FIGURE 3.4   A summary of the possibilities for excitation and deexcitation that do not lead to fluorescence. Short-wavelength light can raise an electron to the S_2 state rather than to S_1. Deexcitation from either state may occur by internal conversion (IC), losing energy as heat, although a large proportion of electrons will simply convert to the S1 state, and then participate in normal fluorescence from there. Electrons that enter a triplet state via intersystem crossing (ISC) are particularly susceptible to chemical reactions, and hence to photobleaching. Alternatively, these electrons cross back to the singlet series or lose their energy somewhat later as delayed fluorescence or phosphorescence.

a higher singlet state than S_1, which is why, in general, we prefer not to excite to these states in fluorescence microscopy. Triplet states tend to be long lasting because the transition back to the singlet state is "forbidden" (which, in chemistry, simply means the transition has a low probability). Various things may happen to the electron after it has entered the triplet state. It may eventually return to the ground state, losing its energy as heat. It may emit a photon but after a much longer than normal time (milliseconds or longer rather than the normal fluorescence lifetime of a few nanoseconds); this is called *phosphorescence*. For a more detailed explanation of these processes, see Anthony et al. (2010).

The consequence of most concern to us is *photobleaching*. Highly excited molecules have, by definition, surplus energy available to undergo chemical reactions. The absorption, fluorescence, and internal conversion processes are usually very fast (approximately nanoseconds), so the molecule typically does not have time to react. However, once a molecule crosses over to the manifold of triplet states, the process of losing the remaining energy is inefficient. Therefore, highly energized molecules live for an extended period of time before they fluoresce. The most important reaction pathway is reaction with O_2, because oxygen has a triplet ground state and triplet–triplet reactions are efficient. Once a molecule has undergone a chemical reaction, the resulting product is probably not going to be fluorescent, so the dye is bleached. Therefore, by removing oxygen we can help prevent photobleaching; most antifade agents are mild reducing agents that scavenge oxygen from the environment. (Many commonly used antifade chemicals, such as paraphenylamine diamine and various gallates, are familiar to photographers, because photographic developers are also mild reducing agents.) An inevitable consequence is that antifade reagents are incompatible with living cells, which require oxygen, so we can only use antifades with fixed material.

The various possible pathways of deexcitation other than fluorescence are summarized in Figure 3.4. In practice, all these reactions occur, so bleaching is an inevitable consequence of fluorescence. The important thing is to pick a fluorochrome in which the reaction we want—fluorescence—is the most likely outcome. The efficiency of fluorescence is called the *quantum yield*, which is simply the ratio of photons reemitted to photons absorbed:

$$\Phi_f = \frac{\# \text{ fluorescence photons}}{\# \text{ absorbed photons}}$$

Quantum yield is an extremely important property for a biological fluorescent probe. The other important "figure of merit" is the *extinction coefficient*, which measures our fluorochrome's efficiency at capturing light:

$$A = \varepsilon c l$$

where

A = absorption strength
ε = extinction coefficient (a property of the molecule)
c = concentration (mol L–1)
l = path length (cm)

A poor quantum yield is undesirable, because it means we will see a lot of bleaching relative to emitted fluorescence. A poor extinction coefficient is not so critical, but we still get better labeling with a high extinction coefficient. Wild-type GFP, for example, has a very good quantum yield but a rather low extinction coefficient, and one of the main aims of engineering new GFP variants was to improve the extinction coefficient (Chapter 12).

WHAT MAKES A MOLECULE FLUORESCENT?

A fluorescent molecule must have highly delocalized electrons, which means that it will have alternating single and double bonds, typically in the form of aromatic ring structures. The larger the conjugated system, the longer the wavelength at which it absorbs. For example, benzene (one ring) absorbs at 260 nm, whereas naphthalene (two fused aromatic rings) absorbs at 320 nm. Figure 3.5 shows us this effect with three common fluorophores. Quinine is excited in the UV and fluoresces blue. The longer, four-ring structure of fluorescein takes its excitation to the blue region and its fluorescence to the green. The addition of two side-chains to the basic fluorescein structure gives us rhodamine, with the excitation further shifted into the green and fluorescence now in the red.

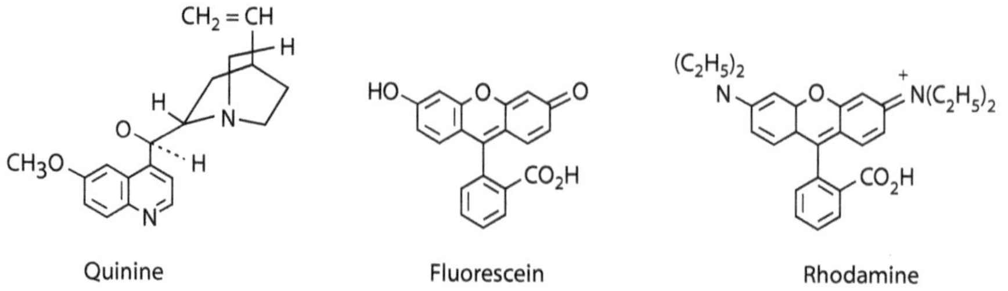

FIGURE 3.5 The molecular structure of three common fluorescent molecules. All have large conjugated systems and hence highly delocalized electrons. The wavelengths of excitation and emission increase from left to right as the molecules get larger.

Microscopy puts certain constraints on the range of wavelengths we can use. If we are to see the fluorescence in an optical microscope, the emitted radiation must be visible: The exciting radiation must have a shorter wavelength than the emitted, so the exciting wavelength is therefore generally short-wavelength visible light (green, blue, or violet) or in the near-UV. Microscope lenses do not transmit well in the ultraviolet, so around 350 nm is the shortest excitation we can normally use. (It is possible to extend this a little by using quartz objectives, and we can capture UV fluorescence on film or with a photomultiplier, and far-red or near-infrared fluorescence with a CCD camera.)

THE FLUORESCENCE MICROSCOPE

OPTICAL ARRANGEMENT

Modern fluorescent microscopes universally use incident-light (epi-)illumination, where the excitation light comes through the objective (Figure 3.6). The system, devised by J.S. Ploem (1967, 1999), has transformed fluorescence microscopy from a complex and dangerous technique, used only by a handful of specialists, to a routine tool. Because the objective lens is also the condenser, Ploem illumination has the incidental benefit of dispensing with the need to set up Köhler illumination.

The major components of any Ploem-based fluorescence system are:

- Light source
- Excitation filter
- Dichroic mirror
- Barrier filter

Each of these components, described in the sections that follow, is an important part of the complete microscope.

Light Source

The key difference between fluorescence and conventional microscopy is that in the fluorescence microscope we illuminate the specimen with one wavelength of light and then look at another, longer wavelength. Therefore, the light must be very bright

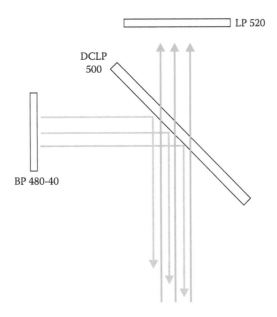

FIGURE 3.6 The basic layout of a fluorescence microscope, based on the design of J.S. Ploem. Incident light is shown in blue and fluorescence in green.

(to excite sufficient fluorescence) and also must have a preponderance of short wave-lengths, because long ones are of no use to us. For this reason, filament lamps are of no real use, and for many years the most common light source was a high-pressure arc lamp. There are probably still more of these around than anything else, but this is likely to change. Such arc lamps typically come in two forms: mercury lamps and xenon lamps.

A *mercury lamp* contains mercury under high pressure while it is in operation (when the lamp is cold, the mercury condenses and there is no pressure in the bulb). Its spectrum is a mix of strong, discrete lines and a continuum background, as shown in Figure 3.7.

There are excellent lines for exciting UV dyes, such as DAPI and Hoechst, and the green lines are perfect for rhodamine and similar dyes. There are also strong lines in the violet and deep blue (indigo), which should get more use than they do on most microscopes. But in the blue region there is relatively little and while the blue output is adequate in most cases, this is the weakest feature of a mercury lamp.

Mercury lamps need careful handling. They operate at a high pressure and temperature, so they are fragile while operating (much less so when cold) and can explode with careless use (wrong voltage, obstructed cooling, or dirt on the lamp). Also, once you switch on a mercury lamp, you must allow it to reach full operation (which takes 10 minutes or so) before you turn it off, and once you turn it off, let it cool down completely before you turn it on again. If you neglect these points, the arc becomes unstable and the lamp flickers.

The traditional rival to the mercury arc lamp has been the *xenon arc lamp*, which has a very strong continuum with relatively minor peaks (Figure 3.8), giving a strong

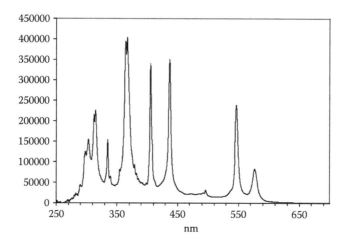

FIGURE 3.7   The spectrum of a high-pressure mercury arc lamp.

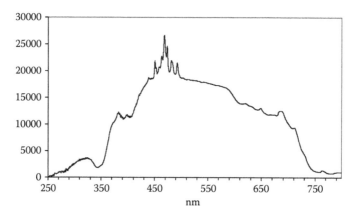

FIGURE 3.8   The spectrum of a xenon arc lamp.

bluish-white light. A xenon arc lamp also offers a longer lifetime than a mercury arc lamp for a similar initial cost, which is an obvious advantage in any lab that uses fluorescence microscopes for many hours a day. However, xenon lamps are weak in the UV region and thus not efficient at exciting dyes, such as DAPI and Hoechst, or calcium indicators, such as Indo. If you do not use these stains, however, xenon is a good choice.

In the past few years metal-halide lamps have become very popular. These are familiar mostly as high-intensity projector lamps and are mercury-based lamps with other metal salts added to give a wider spread of spectral lines. This can tailor the spectrum for almost any requirement, the most obvious being to eliminate the deficiency of a mercury arc in the blue region. Metal-halide lamps have a long lifetime—up to 10 times that of a mercury arc—and are now routinely available as original equipment on most microscopes and also as after-market conversions to replace mercury lamps. They are usually fitted with a liquid light-guide so that they

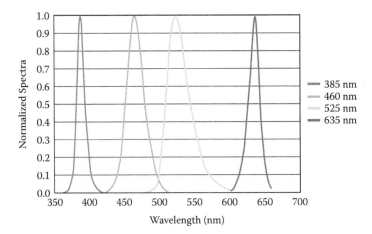

FIGURE 3.9 The spectra of the diodes in the X-cite XLED1 illuminator. (Image courtesy of Kavita Aswani, PhD, Lumen Dynamics, Canada.)

do not need to be mounted directly on the microscope. This allows them to have a cooling fan without introducing vibration to the scope. The light-guides do have a finite life, however.

The most recent development has been the introduction of high-intensity light-emitting diodes (LEDs) as illumination sources. These change the game somewhat, since each LED is one specific wavelength (Figure 3.9). The UV wavelength is a little longer and the blue and green rather shorter than the traditional ones used with mercury lamps (Figure 3.9). This may necessitate filter changes, though in many cases it may be possible to dispense with the excitation filter entirely since only one LED needs to be on at a time. On the other hand LEDs have a very long life (one manufacturer guarantees 20,000 hours or 3 years). The other great benefit is that unlike any other light source for fluorescence they can easily be controlled in intensity, and can be turned on and off as often, and as rapidly, as you wish. Even high-speed pulsed operation is quite straightforward, which makes a lot of interesting experiments possible.

Filter Sets: Excitation Filter, Dichroic Mirror, and Barrier Filter

What made Ploem's system possible was a whole new approach to making filters (Ploem, 1999). Traditional glass filters absorb some wavelengths, while passing others. Interference or dichroic filters reflect some wavelengths, while passing others; these filters can be made extremely specific. They are made of a series of thin film layers deposited precisely (Figure 3.10) and can pass long wavelengths (long pass), short wavelengths (short pass), or a specific band of wavelengths.

Illumination from the mercury (or xenon or halide) lamp passes through an *excitation filter,* which selects the wavelength required to excite the fluorochrome. It is reflected down on to the specimen by a *dichroic* mirror, which acts as a chromatic beamsplitter, reflecting short wavelengths and transmitting long ones. There is therefore little loss of either the illumination or the fluorescence, whereas a half-silvered

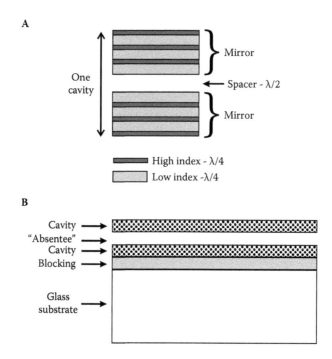

FIGURE 3.10 Diagram of a simple bandpass interference filter. (A) The basic cavity. A Fabry–Perot interferometer made of two multilayer reflectors separated by a half-wavelength spacer that determines the wavelength passed. The layers are alternatively high and low refractive index, each optically ¼λ. (B) The complete filter. There are two identical cavity stacks to sharpen the cutoff. Beneath them may be a blocking filter for unwanted transmissions, an optical colored glass for the same purpose, or both.

mirror would lose half of each, but the dichroic mirror must be tailored to the specific wavelength used.

The objective lens itself acts as the illumination condenser, so the illumination angle will always be the same as the objective numerical aperture. Fluorescent light, which is always of longer wavelength (Stokes' law), returns through the objective lens and passes straight through the dichroic mirror. Beyond the dichroic mirror is a barrier filter to block any residual excitation light. This filter can also be used to separate the light from different fluorochromes. The filter set shown in Figure 3.6 excites in the blue between 480 nm and 490 nm, with a dichroic changing over at 500 nm and a long-pass filter passing all light longer than 520 nm.

Looking at Figure 3.10, these filters might seem almost unimaginably complex, but in fact calculation of the layer thicknesses is straightforward, and the layers are applied automatically by machine. Nevertheless, interference filters are much more expensive than colored glass, and low-cost general-purpose filter sets will often include colored glass filters when possible to reduce total cost.

Filters are normally classified according to whether they pass long wavelengths, short wavelengths, or a band of wavelengths, as shown in Figure 3.11. Figure 3.12 and

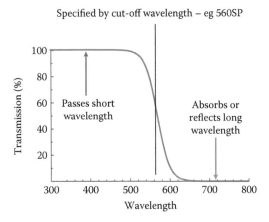

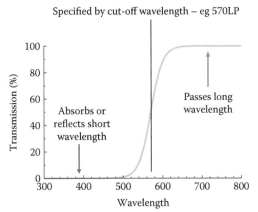

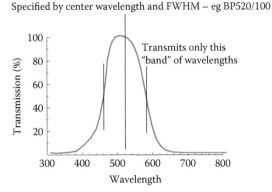

FIGURE 3.11 The three simple types of filter and how each is defined. Short pass (top) and long pass (middle) are defined by their cutoff wavelength, whereas bandpass filters (bottom) are defined by peak transmission and the width of the band halfway between maximum and minimum transmission (full-width half maximum).

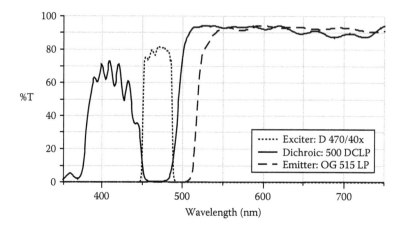

FIGURE 3.12 A typical "low-cost" set for FITC: exciting in the blue from 450 nm to 490 nm, with a dichroic reflecting below 500 nm and transmitting above it, and a simple glass filter transmitting anything longer than 515 nm. Note that the dichroic mirror also transmits around 400 nm, and the excitation filter must be tailored to avoid this region of the spectrum. (Courtesy of Chroma, Inc.)

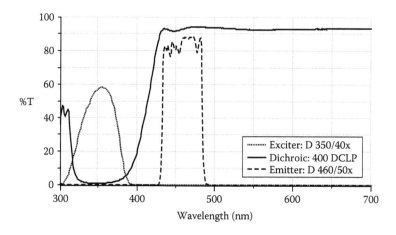

FIGURE 3.13 A typical DAPI or Hoechst filter set using a bandpass filter to exclude any longer wavelength fluorescence (such as FITC), which will also be excited to some extent by the near-UV 360-nm wavelength. (Courtesy of Chroma, Inc.)

Figure 3.13 show the transmission curves of actual filter sets. Figure 3.12 is a low-cost FITC (fluorescein isothiocyanate) set, using a colored glass long-pass barrier filter. All wavelengths longer than 515 nm will be visible in the image. Figure 3.13 is a DAPI filter set with a bandpass interference barrier filter transmitting only in the blue region, so that green, yellow, and red fluorescence will be excluded. With interference coatings it is also possible to make more complex filters that pass multiple wavelength ranges, as shown in Figure 3.14. In this case we can illuminate with several different wavelengths from our mercury lamp at the same time, and pick up

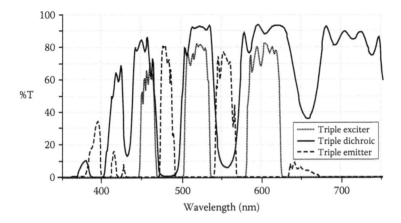

FIGURE 3.14 A triple-labeling filter set, designed for imaging three fluorochromes at the same time. The three-band excitation filter (dotted line) transmits around 390 nm, 480 nm, and 560 nm, and the triple dichroic (solid line) reflects these wavelengths, while transmitting the bands passed by the emission three-band filter (dashed line). (Courtesy of Chroma, Inc.)

two or three different fluorochromes simultaneously. This is convenient from the point of speed, but we will always get the best separation of signals from different fluorochromes, and the greatest sensitivity, by looking at each signal separately.

These profiles of actual filter sets show that, in practice, no individual filter is perfect. Dichroic mirrors are particularly problematic because they cannot include colored-glass blocking layers to compensate for their defects. Therefore, plan your filter sets to work effectively with each other. Figure 3.13 also shows that transmission typically falls off badly at short wavelengths, because the materials used for the coatings absorb in the near-UV. In practice, this does not present a problem with excitation, because a mercury lamp is very strong in the violet and near-UV, but it can be a problem when we want to collect fluorescence in the violet region. Of course, paying a higher price will get a filter made of different materials and offering better transmission.

A lens of high numerical aperture (NA) is important in fluorescence microscopy, because it offers more illumination and better collection of fluorescence, in addition to the best resolution. However, apochromats (Chapter 7) are usually not recommended, even though they offer the highest NA; apochromats contain too much glass, and the special glasses they use often absorb UV. Fluorite lenses are generally best. Be careful, too, with immersion oils, because some are fluorescent themselves.

REFERENCES

Anthony, N., Guo, P., and Berland, K. 2010. Principles of fluorescence for quantitative fluorescence microscopy. In *FLIM Microscopy in Biology and Medicine*, A. Periasamy & R. M. Clegg, eds. Boca Raton, FL: CRC Press, pp. 35–63.

Einstein, A. 1905. Über einen der erzeugung und Verwandlung des Lichtes betreffenden heuristichen Gesichtspunkt. *Annelen der Physik* (Serie 4), 17, 132–148.

Ploem, J.S. 1967. The use of a vertical illuminator with interchangeable dichroic mirrors for fluorescence microscopy with incident light. *Zeitschrift für wissenschaftliche Mikroskopie und mikroskopische Technik*, 68, 129–142.

Ploem, J.S. 1999. Fluorescence microscopy. In *Fluorescent and Luminescent Probes for Biological Activity*, W. T. Mason, ed. New York: Academic Press, pp. 3–13.

4 Image Capture

OPTICAL LAYOUT FOR IMAGE CAPTURE

The optical layout for a microscope presented in Chapter 1 gave us only a virtual image: ideal for observing with the naked eye but which, by definition, cannot be projected onto a screen or to a recording medium (Figure 1.3). To capture the image, we need to take the real image formed by the objective lens and project it as a further real image inside a camera. Figure 4.1 shows how this is done.

In place of the conventional eyepiece, we have what is often (inaccurately) called a "photo eyepiece"; *projection lens* would be a better description. The real image in the microscope tube is farther from the lens than its focus, so it forms a real final image. Usually, this real final image will be magnified and in the past photo eyepieces for film use were typically 3.3× magnification, but digital sensors are smaller than film, so they need a lower magnification.

One problem that we face is focusing: When focusing a microscope in normal use, different people set the virtual image at different distances within the focal range of their eyes, which could be anywhere from 10 cm to infinity! This means that the real image also varies in position, which obviously does not work for a camera, for which the image must be in focus on the film or sensor. In a digital system we can always focus on the monitor, but in practice it is still annoying if the photographic system is not in focus at the same time as the microscope image.

The solution is to put a graticule into the eyepiece. The graticule is located at the "official" plane for the real image formed by the objective, and if the image is in this plane both the graticule and image appear in focus (Figure 4.2). The user must twist the eyepiece focusing adjustment so that the graticule is sharp and adjust the microscope focus so that the image is in sharp focus at the same time. Then, if the camera has been set up correctly, the image will be in focus for the sensor as well.

Exposure control presents another problem. In brightfield or phase contrast, exposing for the average brightness of the field (or the central part of the field) usually gives an adequate result as it does in conventional photography. But in fluorescence (or darkfield), the background is dark (and of no interest). Exposing for the average brightness gives a hugely overexposed result. Modern, so-called smart exposure systems have a fluorescence setting that looks for the brightest points and sets the exposure by these, without reference to the background.

COLOR RECORDING

Our eyes have receptors for three so-called primary colors: red, green, and blue. Other colors of the spectrum are interpreted because they stimulate more than one

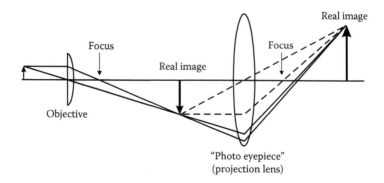

FIGURE 4.1 Optical arrangement in a photomicroscope.

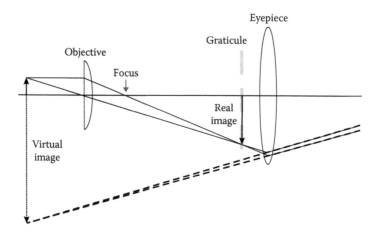

FIGURE 4.2 For photography we need to define the position of the real image formed by the objective (and hence the real image in the camera). This is done by inserting a graticule at the primary image plane to form a reference point.

receptor. The great physicist James Clerk Maxwell (1855) was the first to realize this. Light that stimulates red and green receptors equally appears yellow—and there is no way that our eyes can distinguish between spectral yellow light and a mixture of red and green light. Figure 4.3 shows the primary colors and how they combine to generate all the colors of the visible spectrum. To capture color, therefore, requires three images be captured, one in each primary color. There are two different models for capturing color, and the model that is appropriate depends on how the image is formed when it is viewed.

ADDITIVE COLOR MODEL

Additive color is the simplest to understand, and the easiest example is a data projector. This type of projector forms three different images: one in red, one in green, and

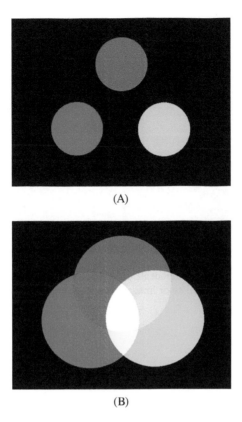

(A)

(B)

FIGURE 4.3   (A) The three primary colors: red, green, and blue. (B) When all three primaries are present in equal amounts we see white, green, and blue create cyan, red and blue create purple, and red and green create yellow.

one in blue. With older model projectors, the three images are obvious because there are three lenses, but most modern projectors still have three images internally. All three images are projected on to the screen in exact (or occasionally not so exact) register. Some projectors use only one image and a rotating filter wheel, so that persistence of vision merges the three colors. On a computer screen the image is made up of tiny red, green, and blue dots placed side by side. Because these dots are below the resolution of the eye, they merge into one image. Figure 4.4 shows the three component primary images that make up the full-color micrograph of mouse skin.

Digital cameras always capture images in additive form, but film (with rare exceptions) and print reproduction use the subtractive method.

Subtractive Color Model

When we print an image on paper (or use normal slide film, where the three images are layered on top of each other), the additive model does not work. Clearly, if we put a layer passing only red light on top of a layer passing only green light, we will

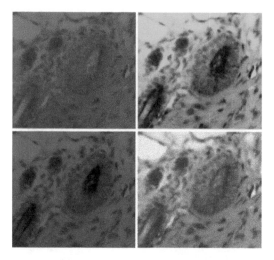

FIGURE 4.4 The three color components of a microscope image (mouse skin, stained with Masson's trichrome) and the full color image formed by combining them.

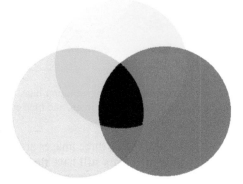

FIGURE 4.5 The three subtractive primary colors: cyan, yellow, and magenta. Where all three overlap everything is subtracted and we have black; where none are present we have white. Where cyan and magenta are present only blue can pass, and so on.

not see yellow. We will see nothing, because no wavelengths can pass through both layers. This is where we need the *subtractive* color model. Instead of adding together the colors we want, we subtract the ones we do not want. White light minus red gives us cyan (green plus blue); white light minus blue gives yellow; and white light minus green gives magenta. These colors, sometimes called the *subtractive primaries*, are shown in Figure 4.5. When no color layers are present we have white light—nothing is subtracted. When all three are present we have black—no light passes. When yellow and magenta are present, both transmit red, so red passes and other colors are subtracted. Cyan and yellow pass green; cyan and magenta pass blue.

CCD CAMERAS

A CCD camera has as its active element an array of a million or more *charge-coupled devices*. A CCD pixel has a photoelectric substance to generate charge when a photon hits and acts as a capacitor to store that charge (Figure 4.6). CCD devices have zero gain, but high quantum efficiency—up to 60% in the orange region of the spectrum (~600 nm) but falling off to half that in the blue. Amplifying such tiny signals hugely magnifies any noise, so high-end cameras for fluorescence microscopy cool the CCD array to minimize random electronic noise (Pawley, 2006).

The CCD elements are tiny—only micrometers across—and the whole detector array is typically only a centimeter or so across. The array is described as a *parallel array* because the charge stored in each element of a row can be shifted in parallel to the next row (Figure 4.7).

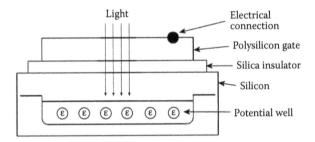

FIGURE 4.6 A single charge-coupled device. Note that the light has to enter the device through the silicon gate, which acts as the read-out connection.

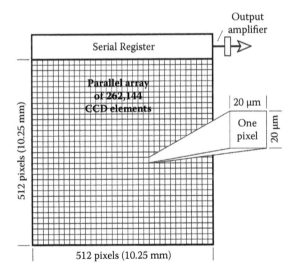

FIGURE 4.7 Sketch of a simple CCD array. Each of the CCD elements shown in Figure 4.11 is only about 20 μm square. At one side of the array is the serial register through which the image is read out.

The pattern of light falling on the array creates different amounts of charge on each detector element. To read each one out, we shift up all the rows in parallel, so that the end row moves into the serial register. We then shift the serial register so that each value, in turn, arrives at the output node. When we have read all the values in the serial register, one by one, we again move everything up one row, and a new row is in the serial register; again, we read this row into the output one element at a time. Figure 4.8 summarizes this process.

Reading out a full-frame device in this way, even at high clock speeds, is relatively time consuming. A shutter is needed; otherwise, light would continue to modify the pixels as they move through the registers, and the image would become smeared. Nevertheless, this readout technique is the method of choice when we need optimum quality. When speed is an issue, two alternative strategies are commonly used: frame transfer and interline transfer.

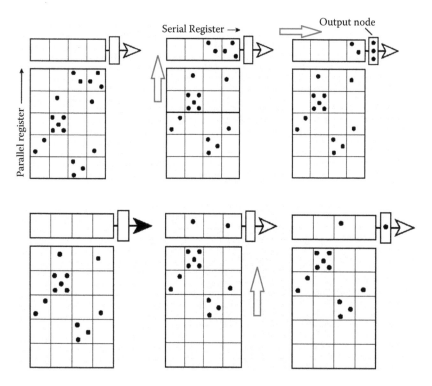

FIGURE 4.8 Reading out a CCD detector. Top row, left to right: All the rows in the parallel register are moved up so that one whole row moves into the serial register, and this is moved sideways step by step so that one pixel is read into the output gate at a time. Second row, left to right: The parallel register is shifted again and the process is repeated for the next row.

FRAME-TRANSFER ARRAY

In a *frame-transfer array*, half the array is masked so that light cannot reach it. Only half the CCD chip is used to acquire the image. At the end of acquisition, we rapidly shift all the registers (in parallel) to the other half. There, the registers can be read out more slowly, while the active part of the array collects the next image. Because the parallel transfer is rapid, we have no need to use a shutter, so we do not waste time or lose any light (a relevant consideration in fluorescence, where any fluorochrome will eventually fade).

Frame transfer gives us speed and sensitivity, but it is expensive and loses resolution.

INTERLINE-TRANSFER ARRAY

An *interline-transfer array* has a transfer line between each row of pixels. After acquiring a frame, we shift everything sideways by one space into the transfer lines. The transfer lines only are then shifted in parallel, row by row, into the serial register and read out, while the active rows collect the next frame. Interline transfer is the fastest possible way to read out a CCD array, but the dead space occupied by the transfer lines considerably reduces sensitivity. The loss of sensitivity does not matter too much in brightfield microscopy but can be a serious problem in fluorescence.

To regain sensitivity without losing the speed advantages of interline transfer, most interline cameras have microlenses on top of each active charge-coupled device, focusing onto it light that otherwise would have landed on the masked area (Figure 4.9). This technique can capture the majority of the light that would otherwise have been lost, giving the device up to 80% of the sensitivity of a full-frame detector. This compromise between speed and sensitivity is probably the most popular for general use in microscopy.

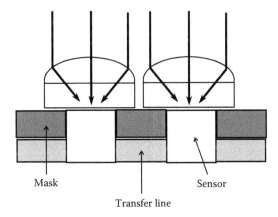

FIGURE 4.9 Microlenses on top of the pixels in an interline transfer camera enable much of the sensitivity lost to the transfer line to be regained.

Back Illumination

The relatively poor blue sensitivity of CCD cameras is caused by absorption in the silicon of the gate structure that overlies each pixel. The back of the device has no gate, so if we thin the silicon down to ~10 μm and illuminate from that side, we can capture more photons, especially at short wavelengths. As you might imagine, the process of back thinning is delicate and has a fairly high failure rate, so back-illuminated CCDs are very expensive. However, for really demanding fluorescence microscopy, where capturing every photon is crucial, they are definitely the way to go.

Binning

Sometimes we are willing to sacrifice resolution for speed and sensitivity, or we have no choice if we are to get an image! We can do this by binning—combining the contents of several elements into one pixel of the output image. Figure 4.10 shows the steps involved in 2 × 2 binning, where each output pixel contains the contents of four array elements and thus has half the resolution of the nonbinned image.

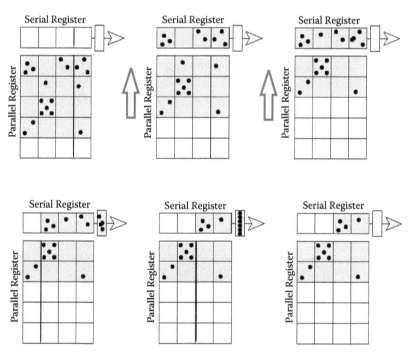

FIGURE 4.10 Binning. (Top row, left to right) We shift up the parallel register twice without shifting the serial register. Now each element in the serial register contains the summed charge of two sensor elements. (Second row) Now we shift the serial register twice before reading the output node. So each value that is finally read out contains the summed charge of four sensor elements.

CAPTURING COLOR

A CCD camera is essentially a monochrome device, and high-sensitivity cameras for dedicated low-light fluorescence are always solely monochrome systems. However, we often want or need to record color, and cameras exist to meet this need, but there are always compromises and trade-offs involved.

Filter Wheels

Filter wheels (Figure 4.11) are the simplest solution to recording color. By rotating a filter wheel in front of the camera, we can capture successive red, green, and blue frames, which we can combine into a full-color image. This system has a lot of advantages. It is cheap and simple, it offers the full resolution of the camera, and the color filters can be of good quality, giving pure separation between the bands. Furthermore, by having a fourth, blank hole or by removing the filter wheel we have a full-resolution, uncompromised monochrome camera.

The big disadvantage of using a filter wheel is speed—it is not usually possible to capture images of moving objects with this system, which means that any live-cell work will probably need to be done in monochrome. A rotating wheel also introduces the possibility of vibration. Nevertheless, for scientific imaging in fluorescence while also being able to photograph stained, fixed material in full color this approach has a lot to offer.

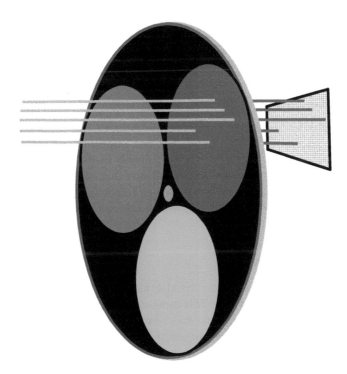

FIGURE 4.11 The filter-wheel approach to color imaging.

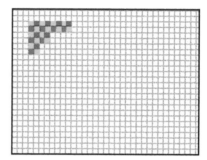

FIGURE 4.12   The Bayer mosaic, used in consumer digital cameras and some scientific ones, has two green pixels for each one in red and blue.

Filter Mosaics

Many cameras in the lower end of the scientific market have a color filter mosaic: each element of the CCD array has a filter element over it. (Most consumer digital cameras also use this approach.) The pattern normally used is known as the Bayer mosaic (Bayer, 1976), which a square contains two green pixels and one each of red and blue (Figure 4.12). This system is cheap, simple, fast, and has no moving parts. However, a mosaic reduces both resolution and sensitivity, and cannot be removed. A computer algorithm ensures that the final image has the same number of pixels as the sensor by supplying the "missing" colors at each point from neighboring pixels. But it is clear that we cannot get true full resolution in three colors from this approach. It also makes binning tricky, because adjacent pixels are different colors; many color cameras therefore switch to monochrome when binning.

Three CCD Elements with Dichroic Beamsplitters

The third solution to recording color is to use three CCD elements with dichroic beamsplitters (Figure 4.13). This approach is fast, offers full resolution, and has no moving parts. This type of construction is commonly used for studio-quality television cameras, but these are much lower resolution than scientific cameras, so that problems of alignment are simplified. With high-resolution, cooled scientific detectors this technique becomes both expensive and complex, so only a few, high-end manufacturers offer this approach.

BOOSTING THE SIGNAL

CCD (and the similar complementary metal-oxide semiconductor [CMOS]) cameras are very sensitive but have no gain. This means that with low signals we are in danger of finding our signal swamped by noise, even in a cooled camera. The problem is particularly intense when we want to image at high speed, following physiological processes. The two possible solutions are either intensifying the image before we capture it (intensified CCD) or amplifying it as we read it out (electron multiplying CCD, or EMCCD).

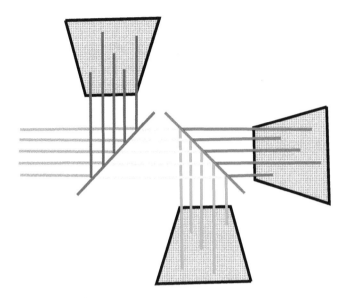

FIGURE 4.13 With two dichroic mirrors, we can capture color with three CCD detectors. This is high resolution but complex, delicate, and expensive.

The major source of noise is the final readout stage, so an EMCCD has an extra multiplying stage to increase the signal before readout. The electrons are moved from the serial register (Figure 4.7), into another register, the multiplying register (Figure 4.14), before the readout stage. In this register the electrons are moved from stage to stage not by the normal low voltage but by a higher voltage between 20 and 40 volts. This is enough to accelerate them slightly, so that sometimes a secondary electron will be produced, adding to the charge in the well. This has a low probability (1%–2%) but after moving through 500 or more stages the signal will be greatly amplified. Since the amplification is stochastic, it will increase noise, but with a weak signal this is much more than offset by the fact that readout noise is now insignificant compared to the amplified signal. The multiplying voltage can be varied or

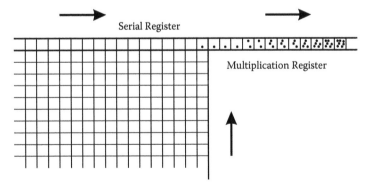

FIGURE 4.14 Electron multiplication. Even though many transfers do not generate any new electrons, in the end the gain is substantial.

turned off so that the camera will also work as a normal CCD where sufficient light is available.

Since speed is always likely to be an issue, EM cameras are frame-transfer devices. EMCCD is a relatively new technology and is not cheap, but has become essential where high speed (as in spinning disk confocals, Chapter 9) or extreme sensitivity (PALM/STORM single molecule imaging; Chapter 17) is required.

Intensified CCD is an older technology and in most applications has been superseded by EMCCD. A *microchannel plate* (MCP) image intensifier is placed in front of the CCD. MCPs will be more familiar in night-vision devices; in fact, if your lab has a multiphoton microscope, you will probably have an MCP-based viewer to see the path of the infrared beam. Essentially an MCP is an array of microscopic photomultiplier tubes (PMTs; Chapter 5) each only about 10 μm across. Figure 4.15 shows the basic layout.

It is not possible to have individual dynodes in such a tiny device, so the plate consists of a high-resistance material with about 100 volts across it. This means

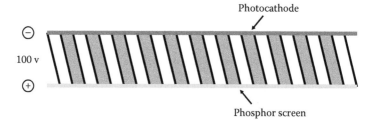

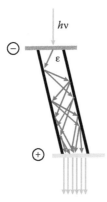

FIGURE 4.15 Layout of a simple microchannel plate image intensifier. (Top) Overview of part of the plate. It is made of high-resistance material so a continuous potential difference will exist between the top (cathode) and the bottom (anode). On top of the plate is a photoelectric substance that will emit electrons when light hits it. At the bottom is a phosphor screen to turn electrons back into light. (Bottom) What happens in one channel. An electron emitted from the photocathode is accelerated down the hole, but since the channel is angled the electron will hit the wall and generate secondary electrons. These in turn will generate more and more electrons on their way down, and at the anode they will stimulate the phosphor to emit light.

there will be a continuous voltage gradient along the length of each channel, and the walls of the channel (which is under vacuum) will act as dynodes. At the top of the plate is a photocathode, just as in a PMT, but at the base we have a phosphor screen that produces light when electrons hit it. A photon hitting the photocathode will (with luck) cause an electron to be emitted. This will hit the wall of the channel somewhere (the channel is angled to make sure that this happens) and generate a shower of secondary electrons. These in turn will hit the wall lower down, and generate more electrons. The electrons are accelerated each time, so by the time they reach the phosphor screen (the anode) one photoelectron has generated thousands of secondaries. These in turn generate thousands of photons as they hit the phosphor. In an intensified CCD camera these are then channeled to the CCD array by a fiber-optic bundle.

MCPs share with PMTs the problem of low quantum efficiency—some photons will fail to generate a photoelectron. Many will not even land over a channel. But, as with a PMT, the gain is then very large. The inherent inefficiency makes it a noisy device, which is why EMCCDs have taken over in most applications. However the MCP has the advantage that it can be switched extremely rapidly. Reversing the voltage to the plate will act as an instant shutter, cutting off the signal, and switching it back will restore it just as quickly. A CCD cannot be switched at such a speed, and a mechanical shutter cannot even come close. This means that intensified cameras are mostly used where high-speed time gating is important.

REFERENCES

Bayer, B.E. 1976. Color imaging array. U.S. Patent 3971065.

Maxwell, J.C. 1855. Experiments on colour, as perceived by the eye, with remarks on colour-blindness. *Transactions of the Royal Society of Edinburgh*, 21(2), 275–298.

Pawley, J.B. 2006. Appendix 3: More than you ever really wanted to know about charge-coupled devices. In *Handbook of Biological Confocal Microscopy*, J.B. Pawley, ed. New York: Springer, pp. 918–931.

5 The Confocal Microscope

THE SCANNING OPTICAL MICROSCOPE

Instead of forming an entire image at one instant, a scanning optical microscope (SOM) scans a beam of light across the specimen in a regular pattern, or *raster,* and forms its image point by point. A point of light is focused to a diffraction-limited spot on the sample by the objective lens. An image is built up point by point by a detector, either below the slide (if the microscope is exclusively an SOM) or below the condenser (which is more practical if the microscope is also used for conventional imaging). SOMs have been used for about 50 years, and the image they produce is formally and in practice equivalent to the image given by a conventional microscope. Why, then, use a scanning microscope? The most common reason is that image processing and analysis are facilitated by having the picture broken up into a series of points (pixels); before the advent of digital cameras, the SOM was the simplest way to achieve this. Scanning also facilitates contrast management when dealing with samples that have either very high or very low contrast. However, what made the SOM an essential part of cell biology was the simple modification that introduced the *confocal* imaging technique. The transformation from SOM to CSM (confocal scanning microscope) is summarized in Figure 5.1.

THE CONFOCAL PRINCIPLE

In a confocal microscope, as in the SOM, a point light source is imaged on the specimen by the objective. This image is no longer a point, but an Airy disk, its size depending on the numerical aperture (NA) of the lens: the larger the NA, the smaller the spot will be, and hence the better the resolution (Chapter 1). This spot is then scanned over the specimen; historically this was done by moving the slide (Davidovits & Egger, 1969), but commercial microscopes normally move the beam of light, as discussed later in this chapter (White et al., 1987). For *confocal* imaging (meaning that there are two coincident focal points), we then collect the light with the objective lens and once again bring it to a focus in front of the detector. At this point, we place a pinhole: in principle, the closest approximation we can make to a geometrical point. Figure 5.1 shows this as an epifluorescence microscope, but by using a half-silvered mirror instead of a dichroic mirror for the beamsplitter, we can use reflected light instead with identical geometry. In either case, the transmitted light can be collected, too, but that will not give a confocal image.

What is the point of doing this? This simple layout has the surprising (but easily understood) property of rejecting information from outside the plane of focus, as Figure 5.2 shows. The light that is in focus (solid lines) is brought to a small spot at

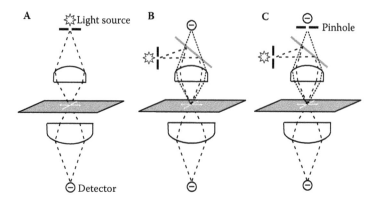

FIGURE 5.1 (A) A simple scanning optical microscope (SOM). A point of light is focused to a diffraction-limited spot on the sample by the objective lens. A transmission image is built up, point by point, with a detector below the condenser, while the slide or the beam is scanned in two directions. (B) The system modified for fluorescence imaging, with one detector recording the fluorescence image and another capturing the transmission image. (C) The addition of a pinhole now turns the microscope in panel B into a CSM, whereas the bottom detector still collects a nonconfocal image.

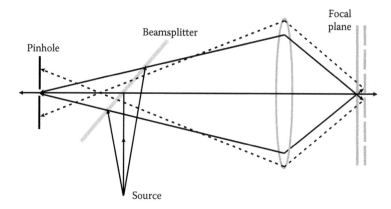

FIGURE 5.2 A basic confocal optical system, showing paths of in-focus and out-of-focus light.

the pinhole, and therefore all goes through to the detector. Light falling on an out-of-focus plane (dotted lines) is brought to a spot in front of (or behind) the plane of the pinhole. At the pinhole, the out-of-focus light spreads over a large area, so that very little of this light passes through the pinhole. The improvement this makes to the fluorescence image of a thick sample is dramatic (Figure 5.3).

When we use a conventional microscope, we are restricted to one plane, the plane of best focus, but this restriction does not apply to a confocal microscope. With a confocal microscope, we can carry out *optical sectioning*: imaging individual planes of a thick object (Figure 5.4). The confocal technique transforms optical microscopy into a fully three-dimensional imaging medium. With a suitable motorized stage, we

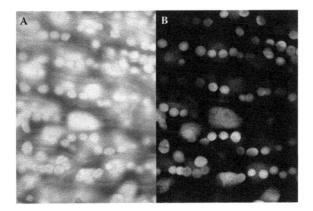

FIGURE 5.3 Fluorescence images of part of a *Selaginella* leaf (chlorophyll autofluorescence). (A) Conventional widefield image. (B) Confocal image. Both images are focused on the same plane.

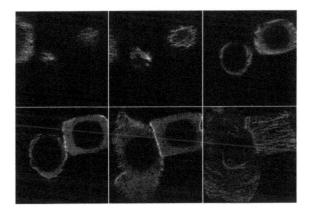

FIGURE 5.4 Six planes from a set of 74 optical sections through an onion root tip squash, showing the cortical microtubule array stained with FITC–anti-α-tubulin.

can automatically collect a complete, three-dimensional data set of our sample. With suitable software, we can then extract information either by resectioning in arbitrary planes or by constructing projections from different views (Chapter 11).

RESOLUTION AND POINT SPREAD FUNCTION

With the optical microscope, we are accustomed to thinking of resolution in the horizontal plane only. As we saw in Chapter 1, that resolution depends directly on the NA of the lens and is given (in fluorescence) by the Rayleigh formula:

$$r = 0.61\lambda/\text{NA}$$

where r is the minimum resolved distance and λ is the wavelength of the light.

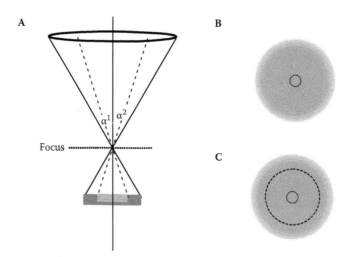

FIGURE 5.5 The size of an out-of-focus spot in relation to numerical aperture. (A) The diameter of the spot at an arbitrary plane depends linearly upon the half-angle of the lens α^1 or α^2, in other words, on the NA. (B) The spot in relation to the pinhole. Even though the vast majority of the light is rejected, some always passes through. It is the ratio of the area of the spot to the area of the pinhole that determines how much is rejected. (C) The spot of the lower NA α^2 superimposed (dotted circle). Because the diameter of each circle depends on the NA, the area—and thus the rejection of out-of-focus light—depends on the square of the NA.

But confocal microscopes also resolve structures in the axial (Z, or vertical) direction, and the factors controlling the resolution in this direction are not quite the same. What controls the resolution in the Z direction is the amount of light rejected by the pinhole. *Out-of-focus light can never be totally rejected.* However large the out-of-focus spot, some proportion of the light will go through the pinhole, as Figure 5.5 shows. So we can never completely exclude out-of-focus objects, and if these objects are very bright, their influence on adjacent planes may be quite noticeable.

What should also be clear from Figure 5.5 is that the amount of the light that is rejected depends on the area of the spot of out-of-focus light compared to the area of the pinhole. So what determines the size of the out-of-focus spot? Looking at Figure 5.2 and Figure 5.5A, we can see that the amount that the light spreads out beyond or before the plane of focus depends directly on the angle, α, at which the rays converge on the focal point—in other words, on the numerical aperture of the lens. Thus, the diameter of the out-of-focus light depends directly on the NA of the lens, and the area of that spot therefore depends on the square of the NA. Although the lateral resolution improves linearly with increasing NA, the axial resolution improves as NA^2. Using a high NA lens is critically important if we want good optical sectioning. For this reason, many manufacturers now make lenses with relatively low magnification and high NA, so that we can get good 3D confocal stacks of reasonably large areas.

Because the axial resolution depends on the square of the NA and lateral resolution is determined by the NA directly, the two are never likely to be the same. In fact, the

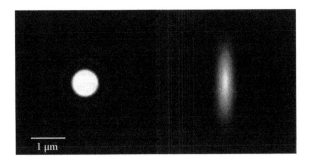

FIGURE 5.6 Horizontal and vertical sections through the confocal PSF of an NA 0.65 lens.

resolution is always worse in the axial direction. How, then, can we summarize the overall resolution of a confocal microscope? A point, imaged on a plane, appears as an Airy disk. The way that same point is imaged in three-dimensional space is called the *point spread function* (PSF), which is an ellipsoid (egg shape; Figure 5.6). Its cross-section through the center is the familiar Airy disk, and a vertical section shows it as an elongated ellipse. The ellipse is at its shortest with the highest NA lenses, where it is a little more than twice its width. With an NA 1.4 lens, we can expect an axial resolution of about 500 nm and a lateral resolution about 200 nm. With NA 0.65, as in Figure 5.6, the axial resolution will be 4 times worse, at ~2 μm.

Although we can apply the Rayleigh criterion in depth and give our resolution that way, in practice it is hard to find a suitable specimen with points directly above each other in a known spacing. So it is common to use a different criterion, the full width at half maximum (FWHM). Fluorescent beads are common and simple test specimens, and if we take an XZ section of one, we see something like the second image of Figure 5.6. We can then, with the microscope software, take a line trace of the intensity vertically through the image and get something like the trace seen in Figure 5.7. The width of this curve, measured halfway between the peak and the background level, is the FWHM. As we saw in Chapter 1, it is a less stringent condition than Rayleigh's (Figure 1.17).

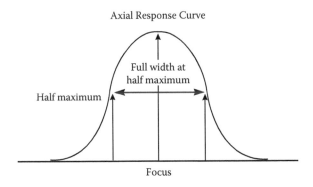

FIGURE 5.7 The definition of full width at half maximum (FWHM).

LATERAL RESOLUTION IN THE CONFOCAL MICROSCOPE

The confocal microscope can also offer a small but useful increase in lateral resolution, especially in reflection imaging (Brakenhoff et al., 1979; McKutchen, 1967; Sheppard & Choudhury, 1977). This is a bit tricky to understand. As the beam scans across a point in the specimen (lower line of Figure 5.8), its Airy disk gradually intersects that point. The illumination of that point is therefore determined by the distribution of intensities in the Airy disk. The edge of the Airy disk illuminates the point weakly, and the image Airy disk (upper line) is therefore dim. As the illuminating spot moves farther onto the point, it illuminates the point more strongly, and the image Airy disk therefore becomes brighter as it moves progressively over the pinhole.

Thus the intensity seen by the detector is a product of the gradually increasing overall intensity of the image Airy disk (resulting from the distribution of intensity in the illuminating Airy disk) and the intensity distribution in the image Airy disk itself. This means that the intensity seen by the detector at any one time is the illuminating intensity multiplied by the imaging intensity. The two curves, multiplied together, give a final curve of intensity versus distance, with much steeper sides (Figure 5.9).

What does this mean? Strictly speaking, we could say that the size of the Airy disk has not changed: The minimum remains in the same position. But the curve of intensity is now much steeper. If we adopt Rayleigh's criterion in its original form (the center of one Airy disk on the first minimum of the next), our resolution has not changed. However, as Figure 1.17 shows, using Rayleigh's criterion on normal Airy disks, the intensity at the point of overlap is 0.735 times the peak intensity. What if we accept the same value with our "squared" Airy disks? In this case, the centers of the disks are closer together, and we get a value for the resolution that is better than

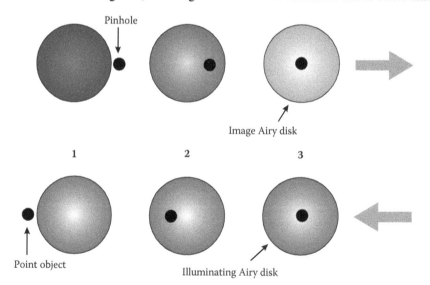

FIGURE 5.8 As the illuminating Airy disk scans across a point, the image Airy disk becomes progressively brighter.

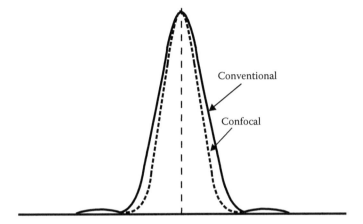

FIGURE 5.9   The intensity distribution in the confocal Airy disk (dashed line) is the square of the intensity distribution in the widefield Airy disk (solid line).

that of a conventional microscope by about $\sqrt{2}$ (1.414). In other words, our minimum resolved distance is now the Rayleigh value divided by $\sqrt{2}$.

We have to be a little cautious about this since the microscope cannot have an actual point detector. If our detector is larger than a point, the improvement is reduced. Fortunately, optical sectioning is not much affected by a finite-sized pinhole, at least until the pinhole size exceeds the Airy disk. In practical terms, therefore, the optical sectioning property of CSM is often more important than its resolution improvement in fluorescence mode, where we usually set the pinhole to be more or less equal to the Airy disk diameter (Cox & Sheppard, 2004). In reflection mode, with more light to play with, we can close the pinhole down and get better lateral resolution.

As far as fluorescence is concerned, optical sectioning is the key factor in any case. Strong fluorescence above and below the plane of focus is the major degrading factor in conventional fluorescence microscopy, and practically useful resolution in biological confocal microscopy will therefore be improved, even if the lateral resolution is unchanged in theory.

As we saw in Figure 5.1C, we can have an additional detector below the specimen, which gives a nonconfocal, scanned image acquired simultaneously with the confocal image. Optical theory tells us that this image will be identical to a widefield image, and we can therefore use contrasting techniques such as phase contrast and differential interference contrast. This can be extremely convenient in practice, because it reveals other cell structures that are not labeled with our fluorescent probes, enabling us to relate labeled structures to the rest of the cell.

PRACTICAL CONFOCAL MICROSCOPES

Figure 5.10 shows a schematic of a conventional laser-scanning confocal microscope (LSCM or CLSM), in which one spot scans the specimen in a regular pattern, or raster. Other implementations of the basic confocal principle have been devised—often sacrificing some degree of confocality in the interest of speed—but these are

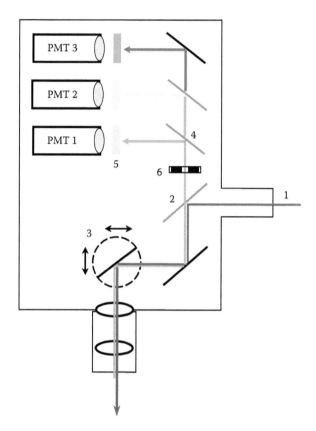

FIGURE 5.10 Diagram of a simple confocal microscope. (1) laser beam entry; (2) primary beamsplitter; (3) scanning mirrors; (4) channel-selecting dichroics; (5) barrier filters; (6) pinhole.

covered in Chapter 9. Commercial confocal microscopes from Leica, Zeiss, Bio-Rad, Nikon, and Olympus all conform (more or less) to this basic plan. (Bio-Rad microscopes are no longer made, but many are still in use.)

The light source is always a laser because nothing else can deliver enough light to one spot. As shown in Figure 5.10, the light from the laser enters the confocal head (1) (either directly or through an optical fiber) and is reflected by the beamsplitter (2) and another mirror down to the scanning mirrors (3), which scan the beam across the specimen, and thence into the microscope itself. The returning fluorescence is descanned (returned to a stationary beam) by the scanning mirrors and passes through the primary dichroic or beamsplitter (2). It is focused to a point at the pinhole (6) and is then split by more dichroic mirrors (4) into two, three, or more channels, each containing a particular wavelength band. Barrier filters (5) further select the detected wavelength range and block any stray laser light before the signal is detected by the photomultiplier tubes (PMTs).

There are various, specific components to consider in such a design, and the choice in each department can make substantial differences to the performance and price of a confocal system:

- The laser or lasers
- The delivery system for the laser light
- The scanning mirrors
- The detection pinhole
- Splitting the signal to different detectors
- Detecting that signal

THE LIGHT SOURCE: LASERS

The word *laser* is an acronym for *light amplification by stimulated emission of radiation*. Lasers amplify light and produce coherent light beams, which can be made extremely intense, highly directional, and very pure in frequency. Electrons in atoms or molecules of a laser medium are first pumped, or energized, to an excited state by an energy source, either light or electric energy. When one electron releases energy as a photon of light, that photon stimulates another to do the same, which in turn stimulates more in a chain reaction, a process known as *stimulated emission*. The photons emitted have a frequency characteristic of the atoms and travel in step with the stimulating photons. The photons move back and forth between two parallel mirrors, triggering further stimulated emissions and amplifying light. The mirrors used must be an integral number of half-wavelengths apart, so that the light continues in phase as it travels back and forth (the total distance between the mirrors can be anything from meters to a few wavelengths but must nevertheless be a precise number of half-wavelengths). Provided that the pump mechanism continually re-excites atoms or molecules that have lost energy, the light builds up into a very powerful beam. The intense, directional, and monochromatic laser light finally leaves through one of the mirrors, which is only partially silvered (Gratton & vandeVen, 2006).

Figure 5.11 sums this up in a generalized diagram. There are many widely different possibilities for the medium that generates the lasing action: solid, liquid, or gas. Most lasers can be assigned to one or the other of the categories of solid state, gas, semiconductor, or dye (liquid). Semiconductor lasers are solid-state devices, of course, but

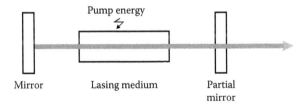

FIGURE 5.11   Schematic of the components of an idealized laser.

a distinction is made because their mode of action is different from solid-state crystal lasers. All except dye lasers are routine components of confocal microscopes.

Gas Lasers

The laser medium of a gas laser can be a pure gas, a mixture of gases, or even metal vapor. A cylindrical glass or quartz tube usually contains the medium. The ends of the tube are set at the angle (the Brewster angle) at which light of one direction of polarization is transmitted without loss. If the tube ends were not set in this way, as *Brewster windows,* reflections from them would interfere with the laser action. With Brewster windows, one direction of polarization is lost and takes no part in the lasing action, but the window is invisible to light polarized in the other direction. The output beam is therefore plane polarized. Two mirrors are located outside the ends of the tube to form the laser cavity (Figure 5.12). The gas lasers used in confocal microscopes are pumped by a high-voltage electric discharge in the low-pressure gas.

Argon lasers are very common in confocal microscopy because their 488 nm line is ideal for exciting fluorescein; their second line, 514 nm (blue–green) was once neglected but is now very useful for yellow fluorescent protein (YFP). More powerful argon lasers offer extra lines—457 nm is very useful for cyan fluorescent protein (CFP) and also aldehyde-induced autofluorescence. Very powerful argon lasers offer a family of close-spaced lines in the near-ultraviolet (UV); these were used in the past in some confocal microscopes, but they introduced substantial technical difficulties and have been superseded by diode lasers.

Krypton lasers have a yellow–green line at 568 nm and a 647 nm red line. These are very useful complements to the argon lines, and some makers add a krypton laser to the standard argon laser. Another alternative is a mixed-gas argon–krypton laser, which offers "white" light (actually a mixture of three discrete wavelengths: 488 nm, 568 nm, and 647 nm). Helium–neon gas lasers are another common alternative, providing 543 nm, 612 nm, and 633 nm lines, although the green 543 nm and orange 612 nm HeNe lasers are rather low in power.

Solid-State Lasers

Until recently, gas lasers were almost universal on confocal microscopes, but semiconductor and solid-state lasers are now taking over and becoming the new standard.

Solid-state lasers are essentially rods of fluorescent crystal pumped by light at the appropriate wavelength to excite the fluorescence. Maiman (1960) made the first

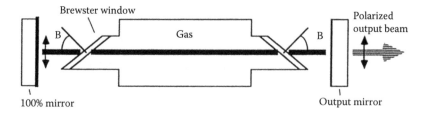

FIGURE 5.12   The components and layout of a gas laser.

laser, a ruby laser powered by a flash tube producing pulses of light at 697 nm, but the most common lasers in confocal microscopes are titanium-doped sapphire crystals and neodymium-doped garnets. The ends of the rod may be two parallel surfaces coated with a highly reflecting nonmetallic film to form the laser mirrors or may be cut at the Brewster angle for use with external mirrors. Solid-state lasers may be operated either in a pulsed manner to generate rapid bursts of light or in continuous-wave (CW) mode. Pumping is normally from other lasers—usually semiconductor diodes, a combination known by the acronym DPSS, standing for diode pumped solid state. Not having other lasers available, Maiman's original laser (and other early lasers) used flash lamps. *Frequency doublers*, special crystals that double the frequency of light (Chapter 8), are often used to turn the 1064 nm wavelength of neo-dymium lasers into the more useful 532 nm (green). These lasers are compact, pow-erful, and low in cost and are now common on confocal microscopes. Doubled DPSS lasers operating at 488 and 561 nm are more recent additions. A 488 nm wavelength can be a direct replacement for argon, and 561 nm is a close match for krypton, a more convenient wavelength than 532 since it permits a wider window for detecting green fluorescence. Both are now very popular.

Semiconductor Lasers

The most compact of lasers, the semiconductor laser, usually consists of a junc-tion between layers of semiconductors with different electrical conducting proper-ties. The laser cavity is confined to the junction region by means of two reflective boundaries. Gallium arsenide and gallium nitride are the typical semiconductors used. Semiconductor lasers are pumped by the direct application of electrical current across the junction, and they can be operated in CW mode with better than 50 per-cent efficiency. Pulsed semiconductor lasers are also available (Chapter 15 deals with their place in confocal microscopy). Common uses for semiconductor lasers include CD and DVD players and laser printers. Red semiconductor lasers operat-ing at around 640 nm and violet ones at 408 nm or 405 nm are routine in confocal microscopes, and blue diodes (470–480 nm) offer another alternative to argon lasers. Also available are 440 nm diode lasers (ideal for CFP) and near-UV 370 nm diode lasers (useful for many dyes, but a challenge optically).

It seems certain that DPSS and diode lasers will take over from gas lasers in the confocal market. The advantages DPSS and diode lasers offer are compact size and much lower heat output into the microscope room. They also have a longer life, but sadly they can still fail. Most manufacturers offer the option of systems with no gas lasers. Diode lasers are typically not as strictly monochromatic as other laser types, and individual lasers can diverge by several nanometers from the nominal wave-length. Although this divergence is rarely a problem in the cell biology, it does mean that more latitude may be needed in filter specification.

Using a laser rather than a lamp means that only certain defined wavelengths are available, and ideally our fluorochromes should be tailored to the laser lines. In practice, the use of lasers means that, in many cases, we are not exciting the fluorochrome opti-mally. Table 5.1 shows the excitation and emission peaks of various common fluorescent stains and how they match with the lines available from argon and argon–krypton lasers.

TABLE 5.1

Common Fluorochromes and Their Compatibility with Popular Laser Lines

Fluorophore	Excitation Peak	Emission Peak	% Max Excitation at			
			488 nm	514 nm	568 nm	647 nm
FITC[a]	496	518	87	30	0	0
TRITC[b]	554	576	10	32	61	1
Lissamine rhodamine	572	590	5	16	92	0
Texas Red	592	610	3	7	45	1
Allophycocyanin	650	661	<1	4	5	95
CY518	649	666	1	<1	11	98

[a] Fluorescein isothiocyanate
[b] Tetramethyl rhodamine isothiocyanate

Careful selection can make a big difference to the effectiveness of our labeling. Thus, among the rhodamine-derivative dyes, TRITC is by far the best choice for a green HeNe laser, whereas Lissamine is excellent with krypton excitation.

Supercontinuum Lasers

Recently commercially available lasers offering "white" light have arrived on the scene, potentially liberating the confocal microscope from the tyranny of fixed laser lines. These use a photonic crystal optical fiber (a fiber with a hexagonal pattern of microscopic tunnels running along its length). When very intense, ultra-short laser pulses are shone into one end, a mixture of nonlinear events takes place in the fiber, generating other wavelengths, so that by the time the light reaches the output end of the fiber an entire spectrum is present. It is not a flat spectrum, so it is not truly white light, but all colors are present.

The actual pulsed laser is usually an optical fiber itself—a fiber doped with a rare-earth element so that it is capable of lasing when pumped by a regular diode laser. These fiber lasers were developed for the telecommunication industry, so they are commonplace items. By arranging a long length of this fiber so that one end shines into the other (a "ring" cavity, though since it is a fiber it is more likely to be a coil) and adding a few other optical components, it can be made to pulse quite spontaneously. Since all the energy of the laser is now concentrated into very short pulses, rather than averaged over time, the light is intense enough for nonlinear interactions to take place in the photonic crystal producing progressively more wavelengths as the pulse propagates. (See Chapter 8 for more on pulsed lasers and nonlinear optical processes.)

We can now pick any wavelength we like to suit any fluorochrome we wish to excite. We can even range through the spectrum and so get the excitation spectrum of an unknown pigment (though we will have to make corrections for the nonuniformity of the excitation). At the time of writing, Leica markets a microscope with a supercontinuum laser, and these capabilities are available. By the time this is published other manufacturers will probably have followed suit.

The major disadvantage is that the intensity at any one wavelength is relatively low, since the power of the laser is now spread across the whole spectrum. This makes it unsuitable for techniques such as fluorescence recovery after photobleaching, or FRAP (Chapter 14).

LASER DELIVERY

In simpler, low-cost, and older LSCMs, the laser line selection is carried out by bandpass filters, shutters, or both, and neutral density filters control the proportion of laser light allowed through. This design is simple, cheap, and effective, but it is slow if we want to collect parallel images with different excitations. Modern high-end systems use an acousto-optical tunable filter (AOTF) to control both line selection and intensity.

An AOTF is a special crystal (tellurite, tellurium oxide), which is birefringent (Chapter 2) and has the unusual property that, if it is deformed by a high-frequency ultrasound signal, its birefringence is reduced for one particular wavelength. Because laser light, as we have seen, is polarized, if the polarization direction is matched to the crystal axis, the ultrasonic vibration changes the direction in which the crystal refracts that one wavelength (Figure 5.13). The frequency of the ultrasound determines the wavelength affected, and the sound's amplitude (volume) determines the proportion of the wave affected. An AOTF can therefore very rapidly change both the laser line and the intensity of the light. Multiple frequencies of ultrasound can be used simultaneously, allowing us to have several different laser wavelengths entering the microscope, each at a different strength.

This high-speed control allows us to carry out many useful tricks. By turning the beam on and off during the scan, we can irradiate a defined shape, as irregular as we like, for bleaching or photoconversion (Chapter 14). We can switch excitation

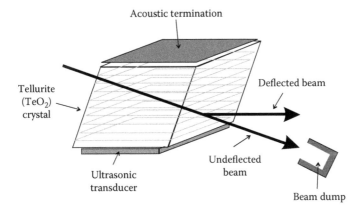

FIGURE 5.13 Diagram of an acousto-optical tunable filter (AOTF). The frequency of the signal generated by the transducer determines which wavelength is deflected and therefore escapes the beam dump and enters the microscope. By varying the strength of the signal, a greater or lesser proportion of the selected wavelength will be deflected.

wavelength on a line-by-line basis to obtain live, two-channel imaging while minimizing bleed-through. Also, the light can be blanked as the beam flies back at the end of the scan line. With most microscopes, no data is collected during this time, so by blocking excitation we lose no signal but substantially reduce bleaching.

The Primary Beamsplitter

Single dichroic mirrors are highly efficient. However, changing dichroics for different laser wavelengths requires that very precise alignment be maintained between them. Also, changing between single dichroics is limiting if we want to collect multiple channels at reasonable speed. One popular solution to this problem is to use a *triple dichroic*, which reflects at three distinct wavelengths and transmits the rest of the spectrum (Figure 5.14; see also Figure 3.13). Because, as Figure 5.14 shows, a triple-dichroic beamsplitter inevitably stops a lot of the returning fluorescent light; several manufacturers have looked for alternatives.

One clever approach has been to use a polarizing beamsplitter, taking advantage of the fact that the laser light is polarized, but (generally speaking) the returning fluorescence is not. This technique is wasteful to some extent: Some incoming light is lost, but there is usually plenty of laser light to spare, and the loss to the returning fluorescence, with a claimed 80% transmittance, is substantially less than with a triple dichroic. However, the beamsplitter is less effective than a dichroic at rejecting reflected laser light, so the barrier filters must be more efficient. Both Bio-Rad and Nikon have used this technique in some of their microscopes.

Leica introduced a system that uses an AOTF to provide a beamsplitter. The company calls this the *acousto-optical beamsplitter* (AOBS). Up to eight wavelengths can be selected with only a few nanometers lost to each. The price for this convenience is considerable complexity; the birefringence of the tellurite crystal splits returning fluorescence into two components of opposite polarization and propagating in different directions. Further, the dispersion of the crystal separates the fluorescence into its component wavelengths. Both effects require compensation with matching (unmodulated) tellurite crystals, so that all wavelengths and polarizations finally emerge as one beam.

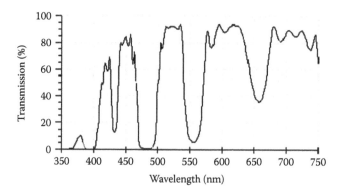

FIGURE 5.14   Transmission curve of a triple dichroic for confocal use (argon-krypton laser) from Chroma. (Courtesy of Chroma Technology Corp.)

Beam Scanning

The beam is scanned by deflecting it with mirrors, normally at a rate between 0.25 and 10 seconds per frame when conventional galvanometer scanners are used. A fast (line) scan in the X direction and a slower (frame) scan in the Y direction are required. The mirror configuration is an important design consideration. Ideally, the mirrors should be light to permit rapid scanning, especially in the X direction. However, there is only one correct spot for the mirror: the plane conjugate with the back focal plane of the objective; only here is the angular movement given to the beam correctly transferred to a spatial movement at the specimen (Chapter 1). Failure to meet this condition creates a risk of *vignetting*: the image is darker at the edges and in the corners, because the whole beam does not enter the pupil of the objective at large scan displacements. Use of a single mirror is thus theoretically ideal; it maximizes efficiency, because there is only one reflecting surface and the mirror can be positioned in the optimal plane to avoid vignetting. However, the mechanical complexity of making a mirror move in two planes increases both cost and the potential for inaccuracy, so that dual-mirror scanning has been the preferred option in most designs.

Some manufacturers simply place two mirrors as close together as possible and hope for the best. A better approach is to introduce optics, forming an image of one of the scanning mirrors onto the other, so that both appear at the same plane to the objective. This can be done with a lens, but Bio-Rad (Figure 5.15) always used two convex mirrors, a relatively bulky system but with very low losses and no risk of chromatic aberration. Although in principle the Y scan mirror, which moves slowly, does not need such a high-performance galvanometer as the X (fast) scan mirror, there are advantages to making the mirrors identical. In this case, the horizontal and vertical scan directions can be interchanged, so that we can have the fast scan either horizontal or vertical. And by cunningly adjusting the scan signal to the mirrors so that a proportion of the voltage going to the X mirror is sent to the Y, and vice versa, the actual scanned frame, or raster, can be rotated without changing the position of either mirror.

Yet another approach is to duplicate the slow scan mirror. By using two linked mirrors for the Y scan, one above and one below the fast scan mirror, a motion is generated that appears to come from the plane of the single X scan mirror. Effectively, the second mirror corrects the errors of the first. This technique provides a theoretically correct scan motion with a minimum of optical elements, but to rotate the direction of scan, you must rotate the whole mirror assembly.

Pinhole and Signal Channel Configurations

The next design consideration is where and how to arrange the pinhole. Typically, we want to separate and detect two or three channels of fluorescence, so we could have one pinhole and split the light passing through it into different channels, or we could split the returning beam of light and then send it to two or three different pinholes, each directly in front of a detector. The first approach places some limits on size; if the splitting takes up too much space, the beam will spread out too far and part will not enter the detector. The second places a premium on alignment; all three pinholes

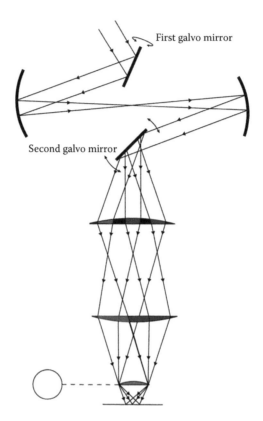

First galvo mirror

Second galvo mirror

FIGURE 5.15 A pair of convex mirrors image one scan mirror on to the other, so that both are congruent with the back focal plane of the objective. Using mirrors eliminates chromatic aberration. This system was used on all Bio-Rad confocal microscopes. (Courtesy of Bio-Rad Microscience.)

must be precisely aligned, and we may have to realign them if we change the beam-splitting for different fluorochromes. However, it does mean that we can set each pinhole precisely relative to the Airy disk size for the relevant wavelength, a refinement that is probably more significant in theory than in practice. There is no consensus on pinhole configuration, and in the end other aspects of the optical design probably dictate the choice: Are we aiming for a compact scanhead, or do we want multiple devices in the detection pathway?

In a simple LSCM design, the size of the Airy disk at the plane of the pinhole is a function of the magnifications of the objective and projector (eyepiece) lenses. A 100× oil-immersion lens of NA 1.3 has an Airy disk with an equivalent diameter of 500 nm at the plane of the specimen. The disk is therefore 50 μm in diameter at the first image plane. A dry lens of 0.7 NA and 40× magnification gives an Airy disk of ~40 μm at the same plane. Some confocal microscopes, such as the Nikon C1, use small, fixed pinholes of these dimensions. Using a 10× projection "eyepiece," the Airy disk becomes ~0.4 nm to 0.5 mm in diameter, simplifying manufacture and alignment at the expense of having more optic elements in the beam path. Making a small pinhole adjustable is another problem, and often pinholes are not round but square, made up of two L-shaped pieces of metal moved by a galvanometer.

Another approach, used in all Bio-Rad microscopes and some Olympus models, is to make the Airy disk much larger—several millimeters—by projecting it to a considerable distance or by using a telephoto lens to enlarge it. Both systems potentially reduce efficiency by introducing additional optical elements, but offer the great advantage that a conventional iris diaphragm can be used as the pinhole. This gives the user extremely flexible control over the size of the confocal aperture.

In recent years, most manufacturers have introduced high-end systems that have abandoned the simple dichroic mirror and barrier filter approach to separating the channels shown in Figure 5.10 (though this remains common on lower-priced systems). The alternative approach is spectral detection: introducing a dispersive device, either a prism or diffraction grating, and directing different parts of the spectrum to different detectors, with no dichroic mirrors or filters involved. This approach allows the user to specify exactly which parts of the spectrum should be recorded in each channel of the image, without being constrained by which filters are installed in the scanhead. It is therefore highly versatile and in principle at least very efficient, because a prism and mirror pass more light than a dichroic and barrier filter. It can also give an emission spectrum, for example, of an unknown fluorochrome. Indeed, it can give a true "spectral image" with a complete spectrum acquired at each pixel.

There are essentially two approaches to spectral detection used in confocal microscopes, which can be regarded as serial and parallel. The *serial* approach, typified by the Leica SP series of instruments, is shown in outline in Figure 5.16. A prism after the pinhole splits the light into a spectrum and movable knife-edge mirrors divert the different parts of the spectrum to different detectors. No part of the spectrum need be lost; one or the other of the detectors can collect every wavelength. In a real system, additional mirrors are used so that we can have the equivalent of "bandpass" collection at each detector, and there may also be more than three detectors. To acquire a spectrum, the mirrors would be moved together to give a fine slit, and then moved in tandem to scan the required spectral range acquiring an image at each position. This generates a lot of data, especially since

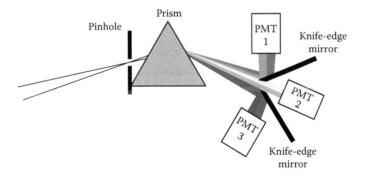

FIGURE 5.16   Spectral detection using movable knife-edge mirrors and single photomultiplier tubes (PMTs).

the Nyquist criterion (Chapter 6) applies to spectra just as much as to images, so if the slit is made 5 nm wide (spectral width, not physical), the step size between images should be 2 nm. Serial collection will, however, give a very high-resolution spectrum, which parallel systems cannot match. The serial approach also offers very high sensitivity, because there are few losses and conventional photomultiplier tubes can be used (see next paragraph). Its disadvantage is that acquiring a full spectral image is slow.

The *parallel* system, shown in Figure 5.17, uses a linear array of detectors, so that the entire spectrum can be collected at once. The Zeiss Meta is an example of this type. Figure 5.17 shows a diffraction grating rather than a prism (of course, either could be used with either system). Diffraction gratings lose more light than a prism but provide a spectrum that is closer to linear. This may be preferable in parallel collection, since the spacing of the detectors in the array is uniform and fixed. The detector is typically a 32-channel photomultiplier: essentially 32 tubes, miniature versions of that shown in Figure 5.18, arranged in a line. Any number of elements can be pooled to give the equivalent of a normal two- or three-channel image, or each can be stored separately to give a full spectral image. This gives a huge speed improvement over serial collection, but with only 32 channels, the spectral resolution is much worse. Some designs seek to overcome this problem by having a variable

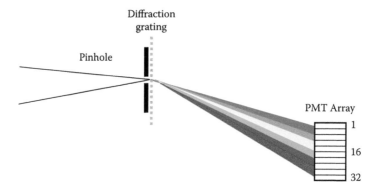

FIGURE 5.17   Spectral detection using a linear PMT array.

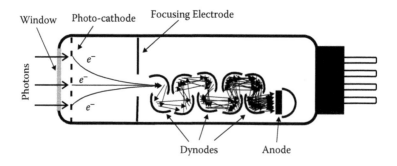

FIGURE 5.18   An end-window PMT.

geometry, so that either the whole spectrum or only part of it lies across the detector array, making it possible to acquire a high-resolution spectrum in several passes. The major disadvantage of parallel detection is that the array detector's sensitivity is significantly worse than that of a conventional photomultiplier tube (PMT). One loses a lot of signal for the sake of speed. This often means that conventional dichroic-based splitting and PMTs must be provided as an alternative within the scanhead, adding significantly to the cost and complexity—and inevitably adding further losses along the way.

Spectral systems of both types also offer utilities for linear unmixing of the spectral data, that is, correction for bleed-through according to known or measured spectra of the fluorochromes. Once the correction has been calculated, from known data or a spectral image, it can be applied to conventional multichannel images, not only spectral ones. This technique gives very good channel separation, even when fluorochromes have quite substantial overlap.

DETECTORS

Detectors for a confocal microscope are typically single elements because they do not have to form an image. The most common type, by far, is the *photomultiplier tube* (PMT). Conventional photodiodes are not sensitive enough for fluorescence detection, although they have been used for transmitted images. *Avalanche photodiodes* have many attractive properties but are very easily damaged by excess light, so at present they are only found in specialist systems such as stimulated emission depletion (STED) microscopes (Chapter 17).

Charge-coupled devices (CCDs), as used in digital cameras (Chapter 4), offer better quantum efficiency (they capture nearly all the photons that hit them) but do not offer any gain, so turning this efficiency into sensitivity is not easy. The amplification necessary for such tiny signals hugely magnifies any noise that is present, so the CCD array in low-light cameras is always cooled to minimize random electronic noise. In fluorescence microscopy, it is normal to acquire an image for several seconds. CCDs have not yet been made in a suitable form for confocal microscope use, because the signal level is too low at any reasonable scan rate. Dwelling a second or more on each pixel would hardly be practicable. Art (2006) gives a detailed discussion of the various possible detectors.

Photomultipliers detect photons arriving at a *photocathode,* which is maintained at a high negative potential (Figure 5.18). This is made of a photoelectric substance, so that it emits electrons when light falls on it. These electrons are accelerated away toward the first *dynode,* which is at a positive potential relative to the cathode. To ensure that the electrons hit the first dynode, they are focused by a *focusing electrode.* This is negative with respect to the cathode, so that electrons approaching it are repelled and follow a curved trajectory. Because these electrons have been accelerated across a potential difference, they have considerable energy when they hit the dynode and each knocks out a shower of low-energy *secondary electrons.* These secondary electrons are accelerated in their turn toward the next dynode, which is at a more positive potential, where further showers of secondary electrons are ejected, and so on through the chain of dynodes. As a result, for each electron that hits the first dynode, thousands or millions travel from the final dynode to the anode (which, for convenience is at earth potential, so that everything else is negative relative to it). The PMT therefore has an extremely high gain, giving a strong signal from a small amount of light. The number of electrons—that is, the current passing between cathode and anode—is proportional to the number of photons arriving over several orders of magnitude, an important feature for quantitative imaging.

Figure 5.18 shows an *end window* tube: The light enters at the end of the tube, and the anode is at the opposite end. Another configuration is the *side window*, in which the light enters at the side of the tube and the dynodes are arranged in a ring around the tube. Both forms are used in confocal microscopes; any difference is more an issue of design convenience than performance.

REFERENCES

Art, J. 2006. Photon detectors for confocal microscopy. In *Handbook of Biological Confocal Microscopy*, J.B. Pawley, ed. New York: Springer, pp. 251–262.

Brakenhoff, G.J., Blom, P., and Barends, P. 1979. Confocal scanning light microscopy with high aperture immersion lenses. *Journal of Microscopy*, 117, 219–232.

Cox, G., and Sheppard, C.J.R. 2004. Practical limits of resolution in confocal and non-linear microscopy. *Microscopy Research and Technique*, 63, 18–22.

Davidovits, P., and Egger, M.D. 1969. Scanning laser microscope. *Nature*, 223, 831.

Gratton, E., and vandeVen, H. 2006. Laser sources for confocal microscopy. In *Handbook of Confocal Microscopy*, J.B. Pawley, ed. New York: Springer, pp. 80–125.

McKutchen, C.W. 1967. Super-resolution in microscopy and the Abbe resolution limit. *Journal of the Optical Society of America*, 57, 1190–1192.

Maiman, T.H. 1960. Stimulated optical radiation in ruby. *Nature*, 187, 493.

Sheppard, C.J., and Choudhury, A. 1977. Image formation in the scanning microscope. *Optica Acta*, 24, 1051–107.

White, J.G., Amos, W.B., and Fordham, M. 1987. An evaluation of confocal versus conventional imaging of biological structures by fluorescence light microscopy. *Journal of Cell Biology*, 105, 1–48.

FURTHER READING

Pawley, J.B. 2006. *Handbook of Biological Confocal Microscopy*, 3rd ed. New York: Springer.

Hibbs, A.R. 2004. *Confocal Microscopy for Biologists*. New York: Kluwer Academic/Plenum.

6 The Digital Image

PIXELS AND VOXELS

In any digital imaging system, whether it is a confocal microscope forming an image with a photomultiplier tube (PMT; Chapter 5) or a wide-field microscope using a CCD camera, we are always dealing with an image made up of individual points: pixels. Often we are also handling samples of a three-dimensional volume: voxels. This quantization has profound effects on our image and how we must treat it. Furthermore, each point (unlike a point in a photograph) can have only certain discrete values. In many cases, the number corresponding to one pixel is an 8-bit value, meaning that it is encoded by eight binary digits (0s and 1s) and can therefore have 1 of 2^8 (256) values; it must lie between 0 and 255.

Some confocal microscopes and most scientific CCD cameras allow 12- or 16-bit image collection, so that each pixel can have 4,096 or 65,536 possible values. This may seem like overkill, because the eye can perceive only 64 or so shades of gray, but it is not really overkill, particularly if we want to do anything numerical, such as ratiometric measurement, with our image. Dividing one 8-bit value by another does not give us a very wide range of possible gray values. Another situation in which we find ourselves running low on grayscales is when we need to take several images for comparative purposes at the same gain and laser settings; the darkest ones will probably have very few tones if we use only 8 bits. Once the number of gray values drops to the point at which the eye can recognize them as separate, the effect is dramatic. Our eyes and our visual processing system have evolved to be very efficient at spotting edges, so our built-in edge detection mechanisms kick in, and all illusion of continuous tone is lost (Figure 6.1). Often the image becomes virtually impossible to interpret. This effect is termed *posterizing* because it resembles the effect of a poster printed with a limited color palette (historically, because of technical limitations in woodblock printing, but now done for dramatic effect).

CONTRAST

The requirements for obtaining an image with maximum information content are not necessarily compatible with the best image quality as judged by eye. It is important that the mean background intensity is always a little above zero. If this is not so, small objects or dim structures may be missed entirely. Even more important, however, is the fact that we cannot know the true boundary of an object. Conventionally, we judge the edge of a structure as the point where the intensity falls to half the maximum, just as we commonly judge resolution as full width at half maximum

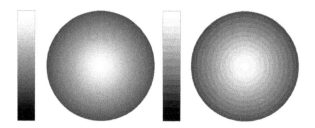

FIGURE 6.1   Posterizing. The eye is very sensitive to edges, so displaying an image that should show a smooth gradation with insufficient gray values makes the image appear to break up into separate objects. Left: 80 gray levels. Right: Only 16 gray levels.

(FWHM). But without knowing the minimum, we will never know where half the maximum lies. Therefore, neither the resolution nor the true size of an object can be assessed.

In an ideal world, the background level should be the dark current from the photomultiplier or CCD, because background fluorescence in the tissue may vary from sample to sample, but in practice you may need to compromise if there is a high level of background fluorescence. Therefore, first remove the sample from the microscope and adjust the black level until a (just) nonzero value is present across the entire field of view. The resulting background will probably be higher than you would choose for visual impact, but this should not be a worry—the image is digital, after all, so we can reset the levels for publication or display. Automatic gain or black level adjustment (or both) can be useful in collecting individual two-dimensional images, but in anything involving time-course or 3D data collection they will be a disaster, because we need all such images to be collected with the same settings.

The intensity at any point in the image must never exceed the maximum that the system can record (except for isolated bits of fluorescent rubbish). If we are storing images in 16-bit form, the maximum possible intensity is approximately 65,000 counts, a value unlikely to be exceeded with any real-life specimen, at least in fluorescence. Twelve-bit storage (4096 levels) is also fairly safe, though extravagant with disk space because the files are actually saved as 16-bit (though no stored values will actually exceed 4095). In many cases, for compatibility reasons or to save space, we store our data as 8-bit images, in which case the maximum recordable intensity is 255. This value is all too easy to exceed, and it is very important to set the gain of the amplifier, PMT voltage, or light intensity to a value at which even your brightest sample will not saturate the image. If the intensity values saturate at any point in an object, then once again, the object's true intensity cannot be measured and neither can its position laterally or in depth. Furthermore, optical sectioning in the confocal microscope, which is essentially only a statistical rejection of out-of-focus light (Chapter 5), will not work properly—an object will be smeared through multiple planes. Confocal microscopes always offer a special color palette in which pixels close to maximum and minimum recordable intensities are shown in contrasting colors, warning when we are at risk of saturation or clipping the black level. This is not so common in CCD cameras, so we may need to take extra care with these.

Another contrast issue that often causes consternation is when an image that looked excellent on the microscope appears very dark when viewed later on a different computer. Digital imaging systems record numerical values that are directly proportional to the amount of light collected. Computer displays, however, whether conventional monitors or flat-panel displays, do not behave in a linear way. Most imaging systems correct for this within their own software, but this does not help when you view the image elsewhere. The relationship between numerical value and displayed intensity is called the gamma, a term borrowed from film-based photography (though the usage is not quite equivalent). High-end imaging software, such as Photoshop or Paint Shop Pro, enables you to set the gamma with which images are displayed, so the images will again look as they did when collected. This software also enables you to change the gamma of the image, which is useful when an image has to be displayed elsewhere, often on a system that is not under your control. Of course, you should change the gamma only on a copy of the original image, because the numerical values will no longer be meaningful in an image saved after you have manipulated the gamma.

SPATIAL SAMPLING: THE NYQUIST CRITERION

In visual light microscopy, we must consider whether we are magnifying enough to make our minimum resolved distance visible to the eye. In digital imaging, we must consider whether we have enough pixels within our minimum resolved distance to record it (Figure 6.2). The minimum is nominally 2.3 pixels within the Rayleigh minimum resolved distance, equivalent to 2 within the FWHM (Chapter 1). This is called the Nyquist criterion (Nyquist, 1928; Shannon, 1949).

When using an NA 1.4 oil-immersion lens, with which we can reasonably expect a resolution of 250 nm or better, our pixel size should not exceed 100 nm, or we will fail to record details resolved in the image. Using a smaller pixel size is termed oversampling. A small amount of oversampling (up to 3 pixels in the minimum resolved distance) can often be justified because such oversampling makes small details easy to see without pixels becoming intrusive. It also allows for a small amount of smoothing to be applied to reduce noise (see next section and Chapter 10) without sacrificing resolution. Substantial oversampling (more than 3 pixels per resolution unit), however, can never be justified. A natural tendency exists to zoom in until the feature of interest fills the field of view, but when we do this we see no more detail, and the fluorescence is likely to fade very quickly because we are bombarding the sample with huge amounts of light.

The Nyquist criterion applies equally in the Z dimension. In confocal microscopy, our depth resolution is substantially worse than the lateral resolution to start with, so undersampling in Z is a bad idea if we want to obtain useful 3D reconstructions from our images. We should regard two slices within the Z-resolution distance as an absolute minimum and preferably take two and a half or three. This may make the resulting image files rather large, but large hard disks and blank DVDs are becoming cheaper every year. The more significant criterion is how long our fluorescence will last. If fading is a problem, reducing the resolution in XY can often be preferable to skimping on Z sampling. A $256 \times 256 \times 128$ data set will be the same number of

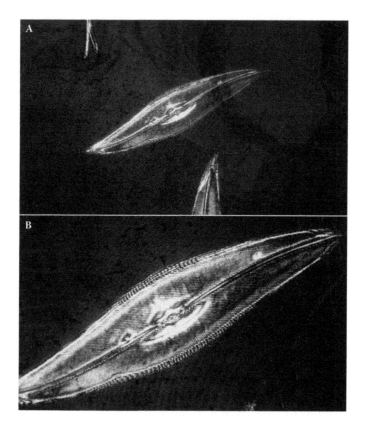

FIGURE 6.2   *Pleurosigma angulatum* frustule. A: Zoom 1.5, pixel size 333 nm. The 500 nm spacing of the holes is not resolved, because there are only 1.5 pixels within the spacing and this is below the Nyquist value. B: Zoom 3, pixel size 166 nm. Now the pattern is resolved.

pixels as a 512 × 512 × 32 data set, but the former generally gives a much better and more useful 3D reconstruction.

It follows that although undersampling—collecting below the Nyquist value—is not generally considered a good thing, in some cases of confocal imaging it can be justified to sample at Nyquist in *Z* and then match the *XY* sampling to that figure. When our fluorochrome would fade, or when a good three-dimensional reconstruction is the prime requirement, this can be a sensible strategy. When our goal is imaging tiny structures such as microtubules, on the other hand, undersampling will not be a good way to go.

TEMPORAL SAMPLING: SIGNAL-TO-NOISE RATIO

It is no use storing 256 (or more) values per pixel if we do not actually make use of this range of values. In the light microscope, we are accustomed to judging an image by its brightness: We regard a good image as one that is bright enough to see clearly.

A digital image can have its brightness adjusted as much as we like; what counts is the ratio of the signal to the random noise in the image, the signal-to-noise ratio.

There are two sources of random noise:

- Electronic noise generated in the system
- Shot noise, which is the random fluctuations in the number of photons coming from our sample

One source of noise will always be *dark current*, the signal produced by our CCD or PMT in the absence of light, resulting from stray electronic fluctuations. With PMTs, a useful strategy for eliminating dark current (when, and only when, we are detecting very dim signals) is photon counting. The idea here is that we set a threshold for detection that represents the number of electrons likely to be generated by a photon. Anything below this threshold we ignore; anything greater is scored as a photon. Of course, if two photons arrive at the same time, they will only be scored as one, which is why this approach is useful only for faint signals. In a PMT, the other source of noise is the fact that the electron multiplication (Chapter 5) is a stochastic event: On average, all electrons liberated from the photocathode are multiplied by the same amount, but individual electrons could vary substantially from the average.

The source of dark current in CCDs is exactly the same as in PMTs—random thermal movements of electrons—but the amplification situation is different. When we read the signal from a PMT it has already been amplified. The signal from a CCD has not and is tiny, so it is very vulnerable to noise at the readout stage. The subsequent amplification is not a stochastic process, so it will not add much extra noise. The common solution is therefore to cool the CCD detector array. The more the detector array is cooled, the better the signal-to-noise ratio we can expect. In astronomy, detectors are often cooled to liquid helium temperatures, but this is rarely necessary in microscopy because other sources of noise are more significant. (Cooling also reduces the dark current in a PMT, but it does not affect multiplicative noise, so again there is little point in doing it.)

Unless something is wrong with the microscope, shot noise will always be the major problem. If we recall (Chapter 1) that the Rayleigh criterion shows us that the contrast between objects at the resolution limit is rather low—only 25% of the contrast range between the object and background—it is easy to see that an inadequate signal-to-noise ratio makes the theoretical resolution of our system a purely academic value (Figure 6.3).

Shot noise applies both to confocal and to wide-field microscopy. The big difference, though, is that when we take a photograph with a one-second exposure, we are collecting data from each point for one second. When we collect a confocal image of 512×512 pixels in a one-second scan, we are collecting data from each point for only four millionths of a second. To get adequate sampling of the signal, we need to use a slower scan time, average a number of frames, or do both. However, we are unlikely to have the patience to collect one frame for the 250,000 seconds we would need to match the collection time of a wide-field image. Wide-field imaging should, therefore, always give us the better image if we are only looking at thin samples and

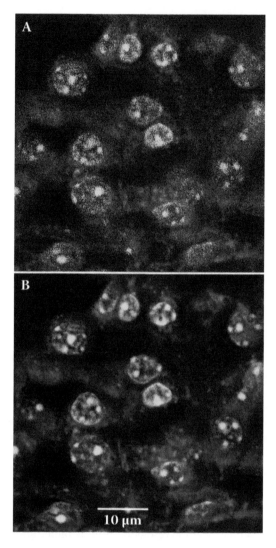

FIGURE 6.3   The importance of signal-to-noise ratio. Mouse kidney stained with DAPI and Alexa-568 phalloidin, imaged at 405 nm for DAPI (blue) and 561 nm for Alexa-568 (red) in a confocal microscope. NA 1.2 lens; pixel size ~100 nm (Nyquist). (A) In a single 1.2 second scan (dwell time 4.56 μs per pixel), shot noise prevents us from making any use of the resolution of our high NA lens. (B) After averaging 10 scans, we get a much more useful image.

do not need to isolate the plane of focus. As in any field of science, choosing the right tool for the experiment is the key to getting good results.

Most digital imaging systems, whether CCD cameras or confocal microscopes, do their initial image collection into a 16-bit buffer (even though they are usually collecting 12-bit data). This provides plenty of space for frame averaging without truncation errors, and we should always average for long enough to give a good signal-to-noise ratio, provided our fluorescence will survive that long. If we are going to do any contrast scaling, it is important that we do it (and get it right) when converting from the 12-bit image to the 8-bit version, because that will populate all 256 gray levels in the 8-bit image. Scaling an 8-bit image will not give us any extra gray levels; it will just spread them out, giving us a histogram with gaps in it. If in doubt, save in 16-bit form. Storage is cheap, but your time is valuable and your specimen may be irreplaceable.

MULTICHANNEL IMAGES

An image of a stained section taken with a color CCD camera of whatever type will be "real" color, stored as a 24-bit image (8 bits in each of red, green, and blue). The camera may also be able to save a 48-bit (3 × 16 bit) image, although even software that can read 16-bit monochrome images tends to give up on 48-bit color. These colors are "real" insofar as they are as close as the camera can get to the specimen's actual colors.

Confocal microscopes use laser illumination and single detectors that record whatever light hits them. The confocal image is monochromatic: We can obtain information about specimen color only indirectly, using different illuminating wavelengths and detector filters. The image is generated one pixel at a time and built up in a frame store; it cannot be viewed directly. A merged image of three detector channels is not the same as a real-color image. We are taking three wavelength ranges—defined by our filters or spectral selections—and putting the whole of each range on the screen in just one primary color. Exactly the same considerations apply in wide-field fluorescence microscopy when we use a monochrome camera and record the image with two or three different filter sets. The green channel, for example, will probably contain colors ranging from blue–green through green to yellow, but we have no way of distinguishing these colors. Even a full-spectral confocal image cannot be viewed in real color on the computer monitor. In this case, the information is present but not in displayable form. In fact, because many confocal microscopes detect in separate channels in the red and far-red, it is quite common to use blue to display the far-red image (Figure 6.4). However, now that 405-nm lasers are common, blue often really means blue, so take care.

Having the image in a computer does provide us with the ability to do all sorts of things with it, from simple averaging to improve a faint image to reconstruction of three-dimensional views from a series of optical sections. Because the displayed image can contain 256 gray levels but the eye can only distinguish 64, it is also common to use a false-color palette to enhance the visibility of fine details. The eye is very sensitive to differences in color and can detect around a quarter million different shades, so this makes all 256 values useful to us (Figure 6.5). Even a gentle tinting can reveal details not visible in the grayscale image, and many confocal microscopes offer a "glow" or "thermal" palette like that of Figure 6.5C as a standard display option. When we display an image in this way, we are not changing any numerical values in our image file, just using a *look-up table* (LUT) to determine

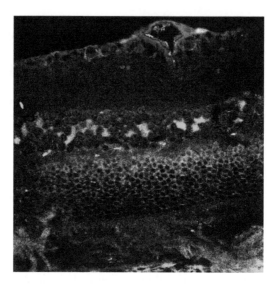

FIGURE 6.4   Three-channel confocal fluorescence image of retina. Green channel is FITC, red channel is CY3, but the blue channel is not a blue dye but TOPRO, a nuclear stain, which fluoresces in the far red. (Courtesy of Jonathan Stone.)

how we display the numerical values on screen. If we export the image to a general-purpose image file format it should be as a paletted file, which writes the LUT as a palette in the file header, leaving the numerical values unchanged. Even so, we should be careful about doing any subsequent manipulations to a paletted file. Saving a paletted file as a compressed JPEG file, for example, will convert it to a 24-bit three-color channel image, and the palette is lost. Simpler manipulations can result in an imaging program's rewriting the palette in a different order and changing the number to match. For safety, do all manipulations in grayscale and only use a palette or LUT when producing the final image for display or publication.

Always remember that the reason something looks green in a confocal or multi-channel image is because we have made it green; a green appearance does not mean the specimen is green in real life. For this reason, always cross-check with the regular image you see through the eyepieces.

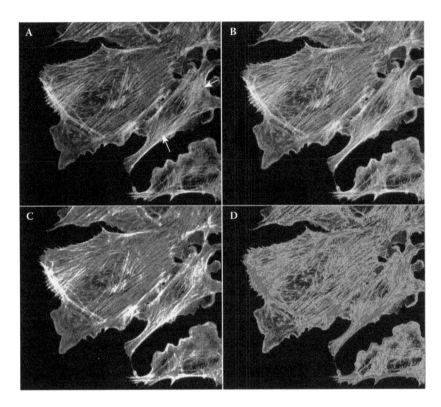

FIGURE 6.5 Confocal fluorescence image of a cultured He-La cell immunostained with FITC against β-tubulin. (A) The original grayscale image. In places where the microtubules are densely clustered it is difficult to make out any details (arrows) because the differences in gray level are close to the limit that the eye can detect. (B) Even a very gentle tinting applied with a false-color LUT can greatly enhance visibility of these details. (C) A "thermal" palette similar to Leica "glow" introduces more color without appearing too false. (D) Sometimes a more striking palette is appropriate. Here contrasting red (bright tones) against green (dim tones). This can be particularly effective for selecting different phases or cell types.

REFERENCES

Nyquist, H. 1928. Certain topics in telegraph transmission theory. *Transactions of the American Institute of Electrical Engineers*, 47, 617–644. (Reprinted as a classic paper in *Proceedings of the Institute of Electrical and Electronics Engineers*, 90, 280–305, 2002)

Shannon, C.E. 1949. Communication in the presence of noise. *Proceedings of the Institute of Radio Engineers*, 37, 10–21. (Reprinted as a classic paper in *Proceedings of the Institute of Electrical and Electronics Engineers*, 86, 447–457, 1998)

FURTHER READING

Pawley, J.B. 2006. Points, pixels and gray levels: digitizing image data. In *Handbook of Biological Confocal Microscopy*, J.B. Pawley, ed. New York: Springer, pp. 59–79.

7 Aberrations and Their Consequences

In Chapter 1 we treated lenses as if they were perfect. Sadly, they are not, and it is precisely for this reason that Antonie van Leeuwenhoek did so much better with his simple microscope than Robert Hooke or anyone else did with compound microscopes of that period. If you cannot correct the defects of your lenses it is better not to have too many of them. As late as the early 19th century the great botanist Robert Brown was making discoveries with a simple microscope—the cell nucleus, Brownian motion—that had eluded the apparently more sophisticated compound microscope. In the next 70 years that all changed, and the microscopes available to biologists improved beyond recognition as the imperfections inherent in simple lenses were finally corrected. Nevertheless, these aberrations, while cured, are still highly relevant to the practical biological microscopist today.

GEOMETRICAL ABERRATIONS

These are intrinsic faults of lenses whose surfaces are portions of spheres, and they can all be constructed geometrically by drawing rays following Snell's law of refraction, which says that when a light ray moves from one medium to another it will change direction according to the formula

$$\sin i/\sin r = n$$

where i and r are the angles made by the light rays to a line normal to the surface and n is the refractive index (Figure 7.1).

In other words, these aberrations are not nearly as complicated as they sound. Using a protractor, compass, and ruler they can all be drawn out on paper so long as the refractive index is known.

SPHERICAL ABERRATION

A lens is relatively flat in the center, but presents a steeper angle to light at the edges. Thus it is more powerful (or has a shorter focal length) at the edges than the center. Rays passing through the edge of the lens (marginal rays) come to a closer focus than those passing through the center (axial rays). Hence the larger the lens aperture the worse spherical aberration will be; in fact spherical aberration becomes worse as the cube of the numerical aperture. Since a lens needs a large aperture to give good resolution it follows that this is a vital correction for a high numerical aperture objective.

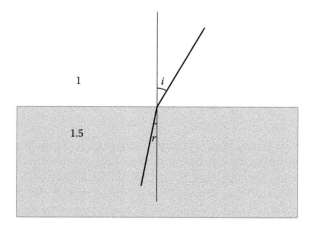

FIGURE 7.1   Snell's law. Rays entering a denser medium are deflected toward a line perpendicular to the surface; rays passing from a denser medium to a less dense one are deflected away from this line.

Spherical aberration (SA) can be minimized by curving the whole lens so that the edges are closer to the specimen than the center (Figure 7.2). This basic approach seems to have been discovered empirically by van Leeuwenhoek and was known to the 18th century astronomer Chester More Hall, but a full correction and a formal treatment of the problem had to wait until the early 19th century when Joseph Jackson Lister (father of the famous surgeon) produced a full analysis and an exact correction (Figure 7.3) (Lister 1830).

As the top picture of Figure 7.2 shows, the way the rays are disposed is rather different on either side of the point of best focus if SA is present. This is shown in detail in Figure 7.4. The vase shape they form is called a *caustic curve* (referring to the fact that it was first recognized when lenses were used as "burning glasses" to light fires, rather than to form images). We can see it directly in the confocal microscope if we take *XZ* images of subresolution beads with spherical aberration present (Figure 7.5).

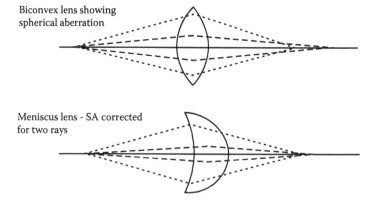

FIGURE 7.2   Spherical aberration reduced with a meniscus lens.

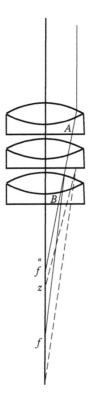

FIGURE 7.3 Lister's original SA corrected lens design, traced from his 1830 paper (Lister, 1830).

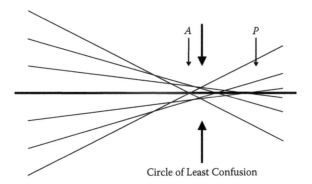

Circle of Least Confusion

FIGURE 7.4 The caustic curve formed around the focus when spherical aberration is present. The sharpest image will be formed at the circle of least confusion. P, paraxial focus; A, abaxial focus.

FIGURE 7.5   Confocal XZ image of subresolution beads showing the shape of the caustic curve.

The best point, known as the *circle of least confusion*, lies some way in front of the focus of the central (*paraxial*) rays and is clearly far from a point, so resolution will not be good. On the side toward the lens, the peripheral (*abaxial*) rays cross the paraxial rays, so there will be a ring of light, whereas on the other side of the circle of least confusion there is quite a concentration of rays around the paraxial focus, surrounded by a broad halo from the abaxial rays.

The effect of this is that if spherical aberration is present the image will look quite different on either side of focus (Figure 7.6). This gives a very easy way of assessing whether spherical aberration is present. Surprisingly enough, it often is present, but the fault lies with the mounting of our sample rather than the objective.

The fundamental difficulty is that spherical aberration correction can only apply to one particular object and image position. Thus a lens can only be corrected for one tube length, and all lenses have their tube length marked on them. In the past a tube length of 160 mm was common on most microscopes, though older Leitz instruments used 170 mm. Petrological microscopes often used longer tube lengths such as 200 mm to make room for polarizers, compensators, and epi-illuminators.

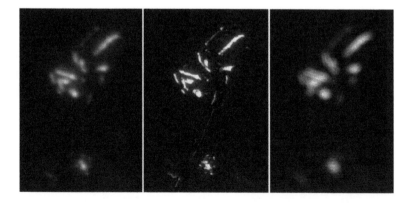

FIGURE 7.6   Spherical aberration has the effect of making the image look different on either side of focus. With no SA, out-of-focus images would look the same on either side of the point of true focus.

In the past 20 years a revolution has taken place and almost all microscopes now use an "infinite" tube length. The objective, on its own, will not form a real image, and therefore the tube can be any length without interfering with the correction for spherical aberration. (In practice, only a few centimeters of leeway are available; since beyond that vignetting—cropping of the edges of the image—will occur.) This makes adding extras such as fluorescence sliders relatively simple. Previously any such additions needed special compensating lenses (which often introduced other problems of their own). To form a real image for the eyepiece an additional lens, called the *tube lens*, is added to the system. This has to be a fixed distance from the eyepiece so the tube cannot be altered above the tube lens.

Likewise, since the refractive index of the coverslip will change the object position for a given image position, it will introduce its own spherical aberration (Figure 7.7). Hence a lens can only be corrected for one specific coverslip thickness, and this is also marked on the lens. It is normally 0.17 mm for a biological objective, or 0 for a petrological or "metallurgical" lens designed for use without a coverslip. This represents the combined thickness of coverslip and mountant between lens and object, and presumes also that the mountant has the same refractive index as glass. Since that is difficult to achieve in practice, lenses of very high numerical aperture are usually fitted with a correction collar to adjust to the coverslip thickness (Figure 7.8). The notional thickness of the different coverslip types is given in Table 7.1, but in practice these are often not accurate, and accurate correction requires adjusting the collar so that the out-of-focus image is identical on both sides of focus.

Correction collars were introduced by Lister and his microscope manufacturer colleague Andrew Ross in 1837 (Figure 7.8). These allowed users to adjust the SA correction for use with or without a coverslip, or with different thicknesses of coverslip. This rapidly became a must-have feature, so some less reputable manufacturers provided a collar that did not actually do anything or even just engraved markings on the lens barrel without anything that turned.

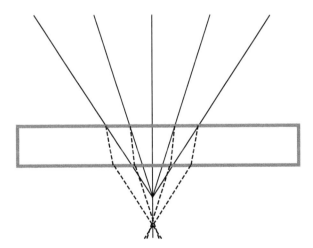

FIGURE 7.7 Introducing a flat piece of glass (coverslip) makes rays that would otherwise have come to a point focus form, instead, the caustic curve of spherical aberration.

FIGURE 7.8 A modern high numerical aperture (NA 0.95) dry lens, fitted with a correction collar to adjust the spherical aberration correction for variations in coverslip thickness. "Plan Apo" tells us that it is flat-field, with apochromatic color correction. "DIC" means the lens is suitable for differential interference contrast, that is, it is strain-free so will not affect the polarization of light passing through it. "M" tells us which DIC prism is correct for that lens. "∞" means that it is corrected for "infinite" tube length; 0.11–0.23 is the range of coverslip thicknesses it can correct for. "WD 0.14" is the free working distance above the coverslip, in millimeters (not a large amount).

TABLE 7.1
Cover Slip Thickness

No.	
0	0.08 mm
1	0.13 mm
1.5	0.17 mm
2	0.19 mm

Oil-immersion lenses (where the initial refraction takes place at one surface only) can be made virtually free of spherical aberration. Since one refractive index applies all the way to the specimen, coverslip thickness will not matter provided that the specimen is mounted in a medium of the same refractive index. Using an oil-immersion lens on a permanently mounted slide should avoid any spherical aberration issues but using the same lens on a live-cell sample in water is guaranteed to cause problems.

As live-cell microscopy has now become more and more important, manufacturers have introduced high numerical aperture objectives designed for water immersion with a coverslip between the lens (in water) and the sample (also in water or

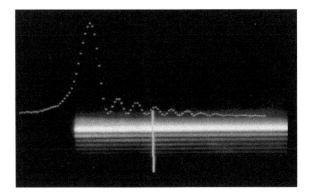

FIGURE 7.9 XZ image of a mirror slide in reflection with spherical aberration present.

saline). Unlike "dipping" lenses, which expect no coverslip and are typically of relatively low numerical aperture (NA), these lenses offer the highest NA possible with water (1.2) and therefore require correction collars to adjust for coverslip thickness. It is essential to spend a little time adjusting this, but once that is done one has the freedom to focus up and down freely without affecting the SA correction.

The effects of SA are often more severe in confocal microscopy than widefield. If our in-focus spot is enlarged by spherical aberration, not only does the image become unsharp, we also will lose a lot of intensity since much of the light will not pass through the pinhole. It is also possible that we will see strange optical effects when some of the bright rings or spots from aberrated, out-of-focus objects happen to hit the pinhole. In the case of layers the image is quite characteristic, with a series of fringes visible in an *XZ* section (Figure 7.9). The problem becomes serious once we start probing deep into a sample with a dry lens, since by definition the SA correction can only be correct at one particular depth, but the whole advantage of a confocal microscope is the ability to image in three dimensions. To make matters worse, we will want to use a high NA lens to get good optical sectioning (Chapter 5). The bottom line is that for a large depth series we should use appropriate immersion media—oil or water, depending on whether our sample is living or fixed.

COMA

Coma is closely related to spherical aberration. If a lens is more powerful at the edges, marginal rays will form a more highly magnified image than the axial rays. Hence the image will be sharp in the center but fuzzy at the edges. Point objects will be smeared into ellipses at the edge of the field (Figure 7.10). This can be distinguished from astigmatism (next paragraph) as the direction of smearing does not change with focus. A lens that is free from both spherical aberration and coma is called *aplanatic*. An oil-immersion objective has a natural aplanatic point (see previous section). Although this position only forms a virtual image, it does give part of the magnification free of aberration, which simplifies subsequent corrections. In general even student-grade oil-immersion lenses will be effectively free of coma as well as SA.

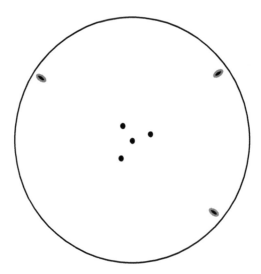

FIGURE 7.10   The effects of coma. Point objects appear as Airy disks in the center of the field but are stretched into fuzzy, radially oriented ellipses at the edge.

ASTIGMATISM

Axial astigmatism—the kind our eyes (and electron microscopes) sometimes have—is not present in a spherical lens. If axial astigmatism is present it can only mean that a lens has been improperly ground, and its surfaces are not portions of spheres, or else that it has subsequently been damaged. (The latter is much the more likely explanation.)

   Radial or oblique astigmatism is, however, an inherent defect of a spherical lens. At the center, the lens presents equal radii of curvature in all directions to arriving light rays, but at the edge the apparent curvature is different in radial and tangential directions. The visual effect is that off-axis points are imaged as radial and tangential lines on either side of focus and as disks at the point of best focus (Figure 7.11). This is not an easy aberration to correct, and even camera lenses (where it is a much greater problem) were not usually corrected for it until the mid-20th century. In microscopy it was often ignored since it is only noticeable at the edges of the field. Even now student-quality objectives often show noticeable astigmatism, though one can assume that research-grade objectives will be fully corrected.

FIELD CURVATURE

Unless a lens is specially corrected, the plane at which objects are in focus is curved not flat (Figure 7.12). Thus when we look at a thin specimen (resin section, blood smear) the edges are not in focus if the center is sharp. Visually this often does not bother us too much when we look down an eyepiece, and it was therefore pretty much ignored by the great 19th century pioneers. The technology to cut very thin sections did not exist so some part of the thickness of the specimen was always in focus, and

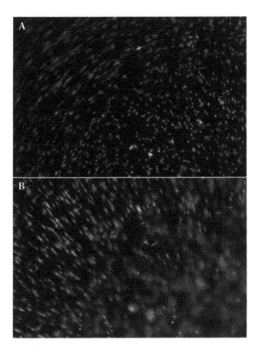

FIGURE 7.11 Images of fluorescent beads, showing the extreme edge of the field from an objective that has radial astigmatism (and a curved field). The center of the field is some way beyond the lower right corner. At one focal position (A) the beads nearer the optic axis (lower left) are in focus, while those farther out (upper right) appear as tangential lines or arcs. At another focus (B) the outer beads are radial lines, but the more central ones just appear out of focus.

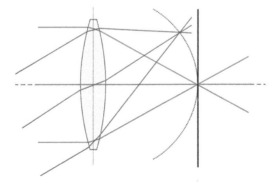

FIGURE 7.12 Curvature of field. The plane of sharp focus is not flat but curved. (Diagram by Ben Frantz Dale, released under Creative Commons.)

images were generally recorded by drawing instead of photography, so one could focus up and down to complete a drawing. The advent of photography provided the first impetus to correct for field curvature, and computer monitors make it almost essential, since the eye does not focus on the center of the monitor in the same way as it concentrates on the center of the field seen through an eyepiece.

Field curvature will also distort any 3D reconstruction of a sample, since we are imaging a curved plane in our sample onto a flat one in the reconstruction. This means that flat field lenses (commonly marked Plan) are pretty much obligatory for confocal imaging.

CHROMATIC ABERRATION

In any real glass the refractive index (n) is not the same for all wavelengths. It is higher for blue light than red; the ratio between the two is called the *dispersion* of that glass. An uncorrected lens will therefore have a shorter focal length for blue light than for red; our image will go blue one side of focus and red the other, and show color fringes at the point of best focus (Figure 7.13). In a conventional microscope this will not matter if we use monochromatic light (e.g., a green filter), but if we want to see color we will have to correct it.

In a confocal microscope chromatic aberration will make no difference to reflection imaging since we are using monochromatic laser light. But it will be quite serious in fluorescence mode since we are scanning the sample with a short wavelength and detecting a longer one. Unless we do something about it our microscope will not be confocal. The spot focused on the pinhole will not be in the same point as the spot of light scanning the sample.

The simplest correction is to combine a diverging lens made from a glass of high dispersion with a converging lens of a low dispersion glass (Figure 7.14), an arrangement first devised by the English amateur astronomer Chester More Hall in 1733. So long as the combined lens is thicker in the center than the edges it will be a positive (magnifying) lens. If we make the overall shape planoconvex we will also minimize spherical aberration. Such a lens is called an achromatic doublet or *achromat* for short.

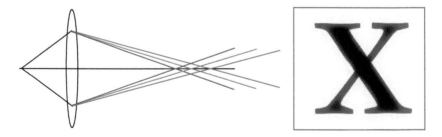

FIGURE 7.13 A simple lens will have a focal length that depends on wavelength, with shorter waves coming to focus closer to the objective. The effect is to give an unsharp image with color fringes (right).

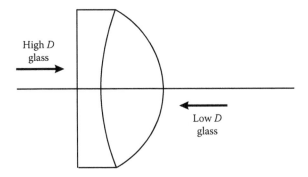

FIGURE 7.14 A simple achromatic lens.

An achromat (Figure 7.15, top) will bring two colors (normally red and blue) to focus at the same point. Other wavelengths will not come to exactly the same focus. We can minimize the spread between the red and blue foci by using the lowest dispersion materials possible. Such lenses are often labeled *fluorite* since fluorite (calcium fluoride) is a material of suitably low dispersion and is often used in these lenses (Figure 7.15, middle).

For the best possible correction we can bring three colors (red, green, blue) to the same focus (Figure 7.15, bottom); such a lens is called an *apochromat*. Apochromats are also designed to correct spherical aberration at two wavelengths. Ernst Abbe was the first to design an apochromat (Figure 7.16A) and the high level of correction it offered made maximal numerical apertures (<0.9 dry and 1.3 in oil) feasible for the first time. However, it contains a lot of glass, which means that its transmission is not as good as a simpler lens, and the special glasses used do not transmit ultraviolet (UV; this is still the case with many modern apochromats).

Abbe's lens also had a very curved field, which is a severe disadvantage for photomicrography. Modern makers therefore produce *planapochromats*, which give a flat field of view. This correction requires a truly amazing number of optical elements (Figure 7.16B), which means that it really was not feasible to manufacture it prior to the development of thin-film antireflection coatings about 50 years ago. Even now we do have to accept that light will be lost in such a complex lens, and there may be cases where simpler fluorite lenses will do as well or better (at a much lower price).

CHROMATIC DIFFERENCE OF MAGNIFICATION

If a lens forms images of different magnifications at different wavelengths it is said to suffer from chromatic difference of magnification or lateral chromatic aberration (Figure 7.17). This means that an image will be sharp in the center, but will show color fringes at the edges. Abbe's original apochromat design has the disadvantage that it introduces chromatic difference of magnification, but fortunately this can easily be corrected by using special eyepieces that are overcorrected, that is, they have the same aberration in the reverse direction. These are called *compensating* eyepieces. Originally they were used only with apochromat objectives, so that changing

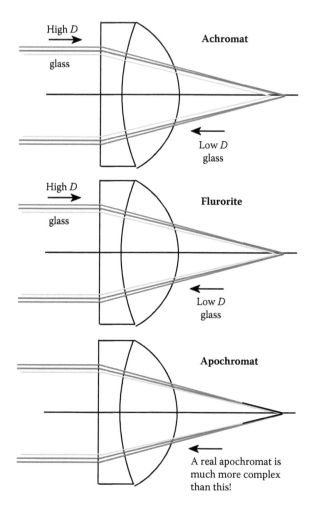

FIGURE 7.15 (Top) An achromat brings red and blue to the same focus; green is not quite the same. (Middle) A fluorite lens narrows the gap between the green focus and the red–blue one. (Bottom) An apochromat (Figures 7.13) brings three wavelengths to the same focus.

objectives might also require a change of eyepieces. This is inconvenient, so from about 1950 most manufacturers designed all objectives, whether achromats, fluorites, or apochromats, to use compensating eyepieces, which made life simpler for both the manufacturer and the user.

With the almost universal adoption of microscopes with infinite tube length, lateral chromatic aberration can alternatively be corrected by the tube lens. Zeiss, for example, does this and therefore no longer uses compensating eyepieces. Olympus still uses compensating eyepieces but does part of the correction with the tube lens. Nikon does all the correction in the objective so that no subsequent compensation is needed. Given this range of possibilities, it will be clear that the objective, tube lens, and eyepiece all need to

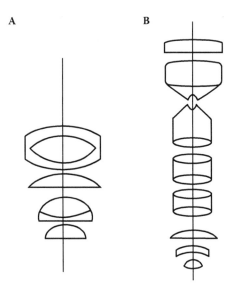

FIGURE 7.16 (A) Abbe's original apochromat design. (B) A modern plan-apochromat. (Drawings courtesy of Gilbert Hartley.)

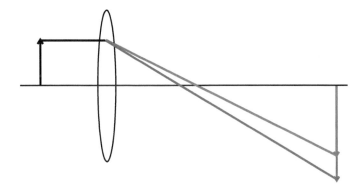

FIGURE 7.17 Chromatic difference of magnification (lateral chromatic aberration). Different wavelengths are brought to sharp focus at the image plane but at different magnifications.

be correctly matched. In a modern microscope you cannot expect to swap objectives and eyepieces between different makes of microscope even if both have infinite tube length.

PRACTICAL CONSEQUENCES

We can generally assume that our microscope, as delivered, is of good quality, but it will not perform at its best unless we specify appropriate lenses. The cheapest (student-grade) lenses a manufacturer supplies are not Plan (flat-field) and will therefore not be suitable for any 3D work. These lenses will also be simple achromats, so their color correction will not be appropriate for confocal fluorescence, though they may be adequate in widefield. When imaging fluorescein isothiocyanate (FITC), for example, we are exciting with blue light and imaging in the green. This is just about the worst possible combination for an achromat.

Plan-fluorites are good general-purpose lenses for widefield and confocal use. The poorer color correction relative to an apochromat is rarely noticeable, but the better transmission often is very useful. Apochromats will give better correction and generally offer higher numerical apertures, which makes for better resolution and optical sectioning, but the sheer amount of glass they contain can absorb light. Also, they typically do not transmit ultraviolet light so they are not suitable for UV excitation in either confocal or conventional microscopy.

Ultraviolet confocal is, in fact, a problem anyway, since no common lenses are corrected into the UV. To overcome this a special correction lens must be put into the path of the UV light to bring its focus to the same point as the visible light focus. An individual correction lens will be needed for each objective that is to be used in the ultraviolet. With the advent of cheap deep-violet lasers most manufacturers have abandoned the idea of ultraviolet excitation in the confocal microscope. Even the violet part of the spectrum is not within the correction range of normal lenses. Since violet lasers (405–408 nm) have recently become very popular on confocal microscopes manufacturers have started to release violet-corrected apochromats, which will perform much better. Since the entire correction range is shifted we might assume that these lenses will perform worse in the far red.

Chromatic difference of magnification will only become a problem if lenses are swapped between microscopes from different manufacturers. The effect then will be color fringes around objects at the edge of the field, and in the confocal we may also see a falling off of illumination at the edges, or vignetting. Once again, this will not be a problem in reflection mode since one is using a single wavelength.

One can take it for granted that spherical aberration is corrected by the lens manufacturer, but the manufacturer has to assume that you will use an appropriate specimen. With a dry lens this means that the thickness of the coverslip and mountant must equal 0.17 mm and that the refractive index of the mountant is the same as that of glass. This is not always easy to achieve, particularly for high NA objectives, since spherical aberration gets worse with the cube of the numerical aperture. High numerical aperture objectives therefore have a compensating collar to adjust for different thicknesses of coverslip, and it is necessary to use it; otherwise you will get a worse result than using a lens of half the price. Even so, with a three-dimensional specimen, spherical aberration will limit the size of 3D stack, which can be collected in a confocal microscope.

This will not be a problem with an oil-immersion lens provided that the specimen is in a mountant of suitable refractive index. The limitation here is just the working distance of the lens, that is, how far you can go before it hits the coverslip. With some lenses this is not far. Glycerol-based antifade mountants are not quite the right refractive index, but in practice one can get away with them in most cases. However if money is plentiful, glycerol-immersion lenses are also available.

Neither of these cases covers the common problem of a living or fixed but hydrated sample in water or saline. For this reason all the major manufacturers have now introduced lenses specially corrected for samples in water but under a coverslip. These are water-immersion lenses but unlike physiological or "dipping" lenses, they have a correction collar to adjust for coverslip thickness. Typically they have a NA of 1.2, about as high as one can get with a specimen in water (remember that an oil-immersion lens does not have an effective NA of 1.4 if your sample is in water). They also have longer working distances than oil lenses. The only drawback is their cost.

The popularity of confocal microscopy has meant that many manufacturers now offer lenses of relatively low magnification but high numerical aperture such as NA 0.5 at 10× and 0.75 at 20×. Since axial resolution is related to the square of the numerical aperture, these will be very useful for obtaining good-quality 3D reconstructions at low magnifications, something that is hard to do with more conventional lenses. Equally, they will be more demanding than standard lenses when it comes to mount thickness and refractive index.

Apparent Depth

Quite apart from the problem of spherical aberration, we run into another difficulty when imaging 3D objects if the lens is not in the same medium as the specimen. The perceived depth will not be the same as the true depth, so that 3D reconstructions will be distorted. If specimen, mountant, and the medium between sample and lens are all equivalent in refractive index no correction is needed. The depth measured by displacement of the slide or objective will be correct. This applies whatever the actual refractive index is, so permanently mounted samples under oil immersion, dry samples in air, and living samples under water immersion will all be fine. This applies with water immersion even if there is a coverslip between the lens and the sample.

With a dry lens and a sample mounted under a coverslip, we will have no problems with spherical aberration (provided that this distance from coverslip to sample is approximately 0.17 mm), but measurements in Z must still be corrected for the "apparent depth" effect caused by the difference between the refractive index of the medium around the lens and the medium containing the sample. Figure 7.18 illustrates this. Rays of light from the object are refracted away from the normal as they leave the coverslip, according to Snell's law of refraction (Figure 7.1). They therefore appear to emanate from the point indicated by the dotted lines. The correct measurement will be given by multiplying the apparent measurement by the refractive index of the mountant (1.5) over that of air (1); in other words the correct figure will be 1.5 times the measured value.

When imaging living cells in water with a dry lens the correction factor becomes 1.3, the refractive index of water. The spherical aberration should be approximately correct about 0.2 mm below the upper surface of the coverslip, and if the lens has a correction collar we can adjust it to correct for the refractive index of water over a wider range of depths. But, where possible, a water immersion lens will still be a better choice.

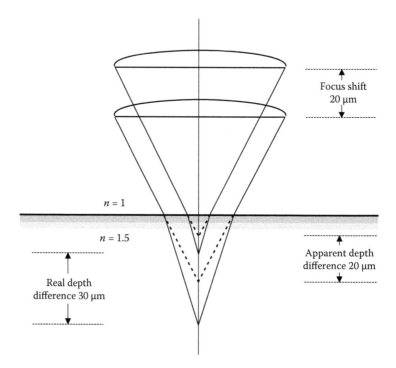

FIGURE 7.18 The relationship between real and measured depth when using a dry lens with a permanently mounted sample.

REFERENCES

Lister, J.J. 1830. On the improvement of achromatic compound microscopes. *Philosophical Transactions of the Royal Society*, 120, 187–200.

FURTHER READING

Fincham, W.H.A. (revised by Freeman, M.H.) 1990. *Optics*, 10th ed. Oxford: Butterworth-Heinemann. (This classic book, first published in 1934, is that very rare thing, an optics textbook comprehensible to the nonspecialist.)

8 Nonlinear Microscopy

MULTIPHOTON MICROSCOPY

Conventional confocal microscopes provide fluorescence excitation in the visible range only. Many fluorochromes used by cell biologists are excited by light in the near-ultraviolet range, including popular calcium indicators (Fura, Indo) and DNA labels (DAPI, Hoechst). To cater to this, some years ago some makers offered confocal microscopes equipped with ultraviolet lasers. These systems suffered from many disadvantages. The lasers then available (high-powered argon) were bulky and troublesome, and not compatible with the optical fibers commonly used to deliver light to the scanning head. The ultraviolet light was toxic to living cells and caused severe bleaching of fluorochromes above and below the plane of focus, limiting the ability of these microscopes to collect three-dimensional and time-course data sets. Furthermore, the lines available were not well suited to calcium ratiometric dyes. Alignment of the (invisible) beam was also tricky and potentially hazardous.

Ultraviolet (UV) lasers also present a major optical problem to the microscope designer. In confocal microscopy, the incident light must be focused precisely on the spot the objective lens is imaging. But microscope lenses are not corrected to have the same foci for ultraviolet and visible light. An ultraviolet confocal system, therefore, requires compensation optics tuned specifically to individual objectives. To a biologist accustomed to having a choice of oil, dry, or water lenses at a wide range of magnifications, this limitation is serious. The UV confocal system was a complex, expensive compromise, and its popularity was limited. When deep-violet diode lasers (which excite DAPI and Hoechst) became available, all manufacturers abandoned true UV confocals. The more recent arrival of semiconductor lasers in the near-UV range reduces the cost and complexity factors associated with water-cooled argon lasers, but the other problems remain, so they have not become part of commercial systems.

PRINCIPLES OF TWO-PHOTON FLUORESCENCE

Two-photon excitation provides a radically different approach to the same problem. This technique overcomes the problems of ultraviolet excitation and provides many new advantages of its own, but of course it requires other trade-offs. To understand the principle of two-photon fluorescence, we need to look at fluorescence itself. *Fluorescence* is the use of light at one wavelength to stimulate the emission of light at another. (The physical chemistry of the process was covered in detail in Chapter 3.) An incoming photon gives its energy to an electron, knocking it into an excited state, a higher energy level that it can occupy only transiently. Soon, in nanoseconds, the electron loses this energy and drops back to its rightful place in the molecule. The energy is emitted as another photon. Because some energy is lost in the process, the photon produced has less energy than the one that originally excited the molecule. Lower energy corresponds to a longer wavelength, so this gives us Stokes' law, which says that the emitted wavelength is longer than the exciting one. The difference in wavelength, known as the *Stokes shift*, varies from one fluorescent dye to another. For conventional fluorescence microscopy, a dye with a large Stokes shift makes it easy to separate the incoming photons from the outgoing ones. (The excess energy is eventually dissipated as heat, but this heat is not usually enough to have any measurable effect on the specimen.)

Two-photon excitation, paradoxically, excites fluorescence with photons of longer wavelength than the emitted light. The laws of physics are not violated because the electron is raised to an excited state by two photons of half the required energy arriving simultaneously. In most cases, the electron ends up in exactly the same (S_1) excited state before it drops to the ground state, so that the fluorescence emitted is identical to that given off by normal, single-photon excitation (Figure 8.1). In this way, we can excite dyes that would normally require ultraviolet with far-red or near-infrared light. However, the excitation pathway is not always identical to the single-photon case, because different selection rules can apply to two-photon excitation. As a rule of thumb, a fluorochrome is excited by two photon events at approximately twice the wavelength required for single-photon excitation, but the two-photon excitation spectrum is often found to be broadened, blue-shifted, or both. In principle, selection rules require that a symmetrical molecule can only be two-photon excited to a different energy state than that reached by single-photon excitation, so the excitation spectra will be quite different (Figure 8.2). The effects of this difference for commonly used fluorochromes are discussed later in the section "Fluorochromes for Multiphoton Microscopy."

The excitation cross-section (i.e., the probability that excitation will occur) does not have any linear relationship with the single-photon excitation cross-section, so a dye that is very effective for single-photon fluorescence may be much less so in a

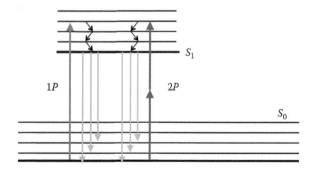

FIGURE 8.1 Jablonski diagram for single-photon and two-photon excitation of fluorescence in the simplest possible case.

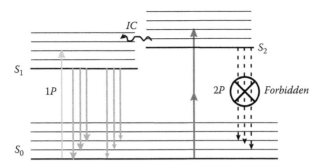

FIGURE 8.2   In the case of a symmetrical molecule, if the $S_0 \rightarrow S_1$ transition is allowed in single-photon excitation, it must be forbidden in multiphoton excitation. Excitation therefore occurs to another state, typically S_2, followed by internal conversion to the S_1 state before fluorescence emission takes place. This changes the shape of the excitation spectrum and may increase susceptibility to bleaching.

two-photon mode (or vice versa). For example, the presence of a triplet state with an energy level at around half the single-photon excitation level can facilitate two-photon excitation (even though the electron does not actually occupy this state in the process). It is likely, therefore, that eventually dyes will be designed specifically for two-photon microscopy; to date, however, only conventional fluorochromes have been used in biological work.

Theory and Practice

The process of two-photon excitation has long been known to physical chemists, and the idea that it could be used in microscopy was first proposed in 1978 (Gannaway & Sheppard, 1978). At that time, experiments were performed on the similar technique of second harmonic microscopy, which is discussed later in this chapter. About 10 years later, a team at Cornell University made the first working two-photon microscope (Denk et al., 1990).

To have two photons arrive simultaneously at a single molecule obviously requires a very large flux of photons impinging on the specimen. Excitation is therefore likely to occur only at the focus of the illuminating light. Above and below this point, the photons are not sufficiently concentrated, and two-photon events are vanishingly rare. In fact, the probability of two photons arriving at a point at the same moment depends on the probability of one photon being there multiplied by the probability of another photon being there. In other words, it depends on the square of the flux density of photons. This dependency has an interesting consequence. If we focus light on a small diffraction-limited spot, the likelihood that two-photon excitation will occur depends on the square of the intensity distribution at the focus. This is exactly the same as the intensity distribution detected in a confocal microscope; in other words, two-photon excitation produces the same imaging properties as confocal microscopy, without any need for a pinhole in front of the detector. With no pinhole, we still see only in-focus structures; optical sectioning in two-photon microscopy is completely analogous to that in confocal microscopy.

The point spread function (PSF) is the same as for a confocal microscope except that the wavelength is longer (typically 700 nm to 1000 nm instead of 450 nm to 650 nm, assuming that we want to excite dyes that require blue or near-UV light in single-photon excitation). The longer wavelength might be expected to reduce the resolution substantially. However, in a confocal microscope a lateral resolution improvement of $\sqrt{2}$ relative to widefield is potentially present (Chapter 3). In fluorescence imaging, this potential essentially cannot be achieved, because we cannot close the pinhole to the tiny size required without losing most of our signal. In multiphoton mode, we do not need a pinhole, so the full $\sqrt{2}$ resolution improvement is always present. So what resolution can we expect?

$$r = 0.61\lambda/\sqrt{2}NA = (0.61 \times 800)/(1.414 \times 1.4) = 247 \text{ nm}$$

This means that using 800 nm excitation we are getting a resolution, in X and Y, which is virtually as good as we will get with confocal excitation in the visible range.

A sufficiently large photon flux continuously applied to the specimen would cook the sample, so the trick is to use a pulsed laser. The instantaneous light intensity is enormous, but the average over time, and hence the heating effect, is comparable to that used in a conventional confocal microscope (CLSM). Typically the pulses are around 100 fs in duration, that is, they are so short they contain only a two-figure number of light waves, (Light travels about 30 μm in 100 fs, so at 800 nm wavelength a pulse will have about 40 waves.) Typical repetition rates are 80 Mhz to 100 Mhz, which means that the time between pulses is 10 ns to 12.5 ns. There are about 200 of these pulses at each pixel as the beam scans the sample.

Lasers for Nonlinear Microscopy

The lasers used to produce these short pulses are usually solid-state titanium–sapphire (Ti-S) lasers delivering pulses in the hundred femtosecond range (1 fs = 10^{-15} s). These lasers are tunable, typically between ~700 nm and ~1000 nm. Tunability obviates another problem of confocal microscopy (especially UV confocal microscopy): the wavelengths available from the laser are never exactly as required. Unfortunately, tuning the laser from one wavelength to another takes time, so it is not possible to image the same field with two different wavelengths in quick succession. Rapid advances are, however, being made in this area. In the past, one had to manipulate knobs and make compensating adjustments when tuning, but now automated Ti-S lasers have computer control for wavelength selection and require no realignment. The laser control software is often incorporated into the microscope software, so that (for example) one can automatically collect an excitation spectrum.

The generation of pulses in a Ti-S laser is essentially passive (though active devices may be added to start the process). As the light passes through the sapphire crystal, the intensity of the electric field modifies the refractive index (the Kerr effect), tending to focus the beam. To encourage this effect, a slit is placed in the optical path, so that focused light passes more easily than nonfocused light. As a result, a single pulse forms in the cavity and travels around it, generating the maximum electric field and therefore the maximum degree of self-focusing. The Ti-S laser (but not its pump laser, of course) works perfectly well with no electrical power. The repetition rate of

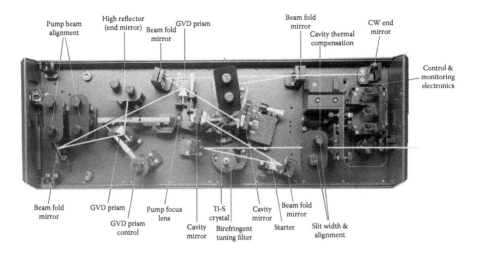

FIGURE 8.3 Beam path in a Coherent Mira titanium–sapphire laser. This is an older-style laser without the complexities of computer control. (Courtesy of Coherent, Inc.)

the laser is the time taken for light to travel around the laser cavity, which explains why the laser is so large and has a complex folded light path (Figure 8.3). The actual Ti-S crystal is quite small.

A Ti-S laser must be pumped with another laser of substantial power. Typically, the nominal efficiency will be around 10%, so that a 500 mW laser requires 5 W to pump it. Solid-state neodymium lasers are used for pumping. The output from these lasers is in the infrared, at ~1050 nm, and is then doubled in frequency with a lithium beryllium oxide (LBO) crystal, which generates a strong second harmonic at 525 nm. These lasers are small enough to fit into the casing of the Ti-S laser in modern "one box" systems. The neodymium laser itself needs to be pumped with red light from two or more powerful diode lasers, which are always mounted externally and coupled by an array of optical fibers.

The pulses of a Ti-S laser are only a few wavelengths long, so they are not mono-chromatic by the standards of continuous wave lasers. In fact, it is possible to esti-mate the pulse length by measuring the spread of wavelengths present. This spread presents a problem when the laser pulses must travel through a dispersive medium, as in optical components. The (slightly) different wavelengths travel at different speeds, and the pulses spread out.

Coupling the laser to the microscope with an optical fiber, as is done in most confocal microscopes, therefore spreads the pulses by dispersion so that they emerge with a substantially longer duration. In the past, several manufacturers have experi-mented with fiber-optic coupling, but all have now abandoned the attempt. Keeping the bulky Ti-S laser in line with the microscope requires both to be mounted on an optical table, adding to the complexity and size of the system. Even the microscope objective, which contains a large thickness of glass, causes dispersion that length-ens the pulses, so the actual pulse length at the sample is likely to be longer than that leaving the laser. To compensate for this some systems are now equipped with "pre-chirp" optics that advance the short wavelengths, so that subsequent dispersion will retard them and restore the original pulse shape.

Advantages of Two-Photon Excitation

Two-photon imaging offers several clear advantages over single-photon imaging:

- Because two-photon events can take place only at the focus of the illuminating light, bleaching above and below the plane of focus is eliminated. In a conventional confocal system (although we are only forming an image of the focused spot formed by the laser beam), the cone of light above and below this spot is also capable of exciting—and bleaching—our fluorescent label.
- Penetration of the exciting light into tissue is increased by at least a factor of two due to its longer wavelength. Scattering of light in a turbid medium is directly related to wavelength (which is why the sky is blue), so longer wavelengths penetrate much better into cells. There is a counterbalancing effect—as we move into the infrared, water starts to absorb light—but in the range between 700 nm and 1200 nm, the gain from reduced scattering exceeds the loss due to absorption. Of course, the ability of the fluorescence to get back out of the tissue is unchanged, but we can now collect even scattered light from the fluorescence, because it no longer needs to come to a focus. Overall, the depth from which we can get useful images is substantially increased (Figure 8.4).
- The exciting light is within the range in which microscope lenses perform efficiently, eliminating or simplifying optical corrections. Microscope lenses are generally designed for the range of ~700 nm to 400 nm, so that UV excitation at 350 nm lies outside the chromatic correction of the objective, but two-photon excitation at 700 nm is just about inside it. Even when we drop into the infrared region, many lenses transmit better than they do in the UV, and the fact that the focused spot might no longer be imaged on the confocal pinhole does not matter so much since …

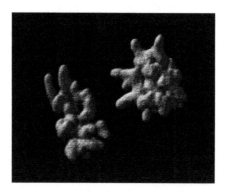

FIGURE 8.4   Meiosis in a living *Agapanthus* anther imaged at a depth of 200 μm. Three-dimensional reconstruction from serial optical sections (see Feijó & Cox, 2001).

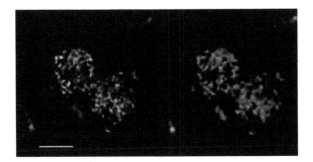

FIGURE 8.5 Mitochondria in a cultured nerve cell labeled with the potential-indicating dye JC1. Green fluorescence predominates where the membrane is less polarized, orange–red in regions of high membrane potential. Scale bar = 10 μm. (Left) Red and green channels. (Right) Ratiometric image showing the ratio of the two channels. (From Dedov, V.N., Roufogalis, B.D., and Cox, G.C., 2001, *Micron*, 32, 653–660. With permission.)

- There is no need for a confocal pinhole. Because two-photon events occur only at the focus of the illuminating beam, two-photon microscopy is depth-selective even with a widefield detector. In fact, chromatic correction becomes almost irrelevant in multiphoton because we need to bring only one wavelength to a focus.
- Long-wavelength light is much less damaging to living cells than ultraviolet or even blue light. Many experiments have shown extended cell viability during multiphoton imaging compared to equivalent single-photon imaging conditions. In the example shown in Figure 8.5, although the JC1 dye is easily imaged in single-photon mode, cell viability was severely limited, making it impossible to complete a useful experiment. Under two-photon excitation, the cells remained viable for many minutes, quite long enough for the experiment.

CONSTRUCTION OF A MULTIPHOTON MICROSCOPE

Multiphoton microscopes from the major manufacturers use conventional confocal microscopes as their basis (even though, in principle, doing so is not necessary). The scanning arrangement is therefore that of a confocal microscope (Chapter 4), and a normal pinhole arrangement is still present, though it can be opened fully for multiphoton imaging. However, this does not make best use of the technique. A key benefit of multiphoton imaging is that scattered light that would not reach the pinhole in a confocal system can still be collected, and to do this optimally, we need detectors to catch this light. Multiphoton microscopes are therefore equipped with *non-descanned* (i.e., widefield) detectors. These can be placed either behind the objective, where the widefield fluorescence illuminator would be, or beyond the condenser, where the transmitted illuminator would be. Of course, to get the best of all possible worlds, we would do both (Figure 8.6). The choice of geometry is particularly relevant if we also plan to do second harmonic microscopy (see later section), because fluorescence is transmitted equally in all directions, but the second

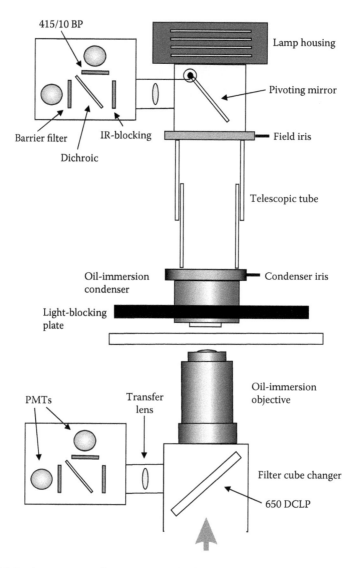

FIGURE 8.6 Arrangement of non-descanned (widefield) detectors in an inverted multiphoton microscope.

harmonic is preferentially transmitted in the forward direction. Because the confocal optics add to the expense and complexity, several smaller firms now make multiphoton-only microscopes, with no pinhole or descanned detection.

FLUOROCHROMES FOR MULTIPHOTON MICROSCOPY

The simplistic analogy that two-photon excitation is exactly equivalent to excitation with one photon of twice the energy is not entirely accurate, as we have seen. Often, the two-photon excitation spectrum is both broader and blue-shifted relative to twice the single-photon spectrum (Xu & Webb, 1996; Xu et al., 1996). The selection rules for the two processes can be quite different; in particular, for centrosymmetric molecules, 1PA and 2PA must occur to different states.

This means that the optimum wavelength for excitation will not be twice the single-photon wavelength, as Figure 8.2 shows, but it also means that, as always when excitation is not occurring to the S_1 state, the rate of bleaching increases. So dyes that are excellent for single-photon fluorescence may not work as well in two-photon fluorescence and vice versa.

The popular fluorochromes fluorescein and rhodamine are symmetrical molecules, and their two-photon excitation spectra are very different from their single-photon spectra (Figure 8.7A, B). It is clear from these spectra that excitation between 740 nm and 810 nm for fluorescein and around 840 nm for rhodamine access a higher excited state with increased likelihood of bleaching, and practical experience bears this out. On the other hand, asymmetrical molecules, such as lucifer yellow and the popular calcium indicator Fura 2, have two-photon excitation spectra that closely match their single-photon spectra at twice the wavelength (Figure 8.7C, D). Green fluorescent protein (GFP) and yellow fluorescent protein (YFP) have similar spectra for two-photon excitation as in single-photon but with a small blue shift (Figure 8.7D, E).

SECOND HARMONIC MICROSCOPY

Second harmonic generation (SHG) is an older microscopy technique than two-photon fluorescence (Gannaway & Sheppard, 1978), but for many years it was overshadowed by the two-photon technique. As electromagnetic radiation propagates through matter, its electric field exerts forces on the sample's internal charge distribution. The consequent redistribution of charge generates an additional field component, referred to as the *electric polarization* (P), which has both linear and nonlinear components. The nonlinear components become significant only at very high light intensities. Their effect is to generate overtones, or *harmonics*, at exact multiples of the original frequency. As far as the second harmonic is concerned, this generation happens only if the molecule's electrical properties are different in each direction, so the ability to generate second harmonics is peculiar to molecules that are not centrosymmetric.

Table 8.1 sets out the key similarities and differences between second harmonic microscopy and two-photon microscopy.

Serious biological use of SHG microscopy has been a recent phenomenon, with the first papers appearing post-2000 (Campagnola et al., 2001; Cox et al., 2003; Zipfel et al., 2003). Since then, this technique has been used largely to image collagen, but SHG has also been used to image styryl potentiometric dyes both in the cuvette and in the microscope, and thus to measure membrane potential in vivo (Campagnola et al., 2001). Recent dye developments have made this a very promising technique in neurobiology (Dombeck et al., 2005).

Collagen fibrils have a triple-helical structure and are both noncentrosymmetric and highly crystalline. It has been known for 20 years that this structure makes collagen able to generate the second harmonic from a wide range of wavelengths in the

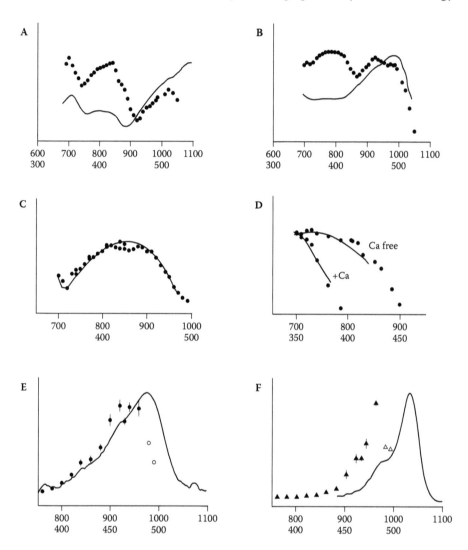

FIGURE 8.7 Excitation spectra for several fluorescent probes in single photon (solid lines, lower figures) and multiphoton (symbols, upper figures). (A) Rhodamine and (B) fluorescein are symmetrical dyes and show very different spectra in the two modes. The nonsymmetrical dyes (C) lucifer yellow and (D) Fura 2 have two-photon excitation spectra that closely match their single-photon spectra (at twice the wavelength). (E) Enhanced GFP (eGFP) and (F) enhanced YFP (eYFP) both show similarly shaped spectra in the two modes but the two-photon spectra are noticeably blue-shifted.

infrared region (Roth and Freund, 1981). Collagen is an extracellular structural protein of major importance to all vertebrates, being a major component of bone, cartilage, skin, interstitial tissues, and basal laminae. Given the importance of collagen in the vertebrate body and the strength of the SHG signal it produces, it is not surprising that collagen imaging has so far been the major use for second harmonic microscopy

TABLE 8.1
Comparison between Second Harmonic Generation
and Two-Photon Fluorescence

Second Harmonic Generation	Two-Photon Fluorescence
Exactly double original frequency	Spectrum of frequencies <2× original
Largely frequency independent	Strongly frequency dependent
Virtually instantaneous: ~1 fs	Lifetime in ns
Propagated forward	Propagated all directions
Coherent (in phase with exciting light)	Incoherent
No energy loss or damage	Always energy loss and associated damage
Requires short laser pulses	Requires short laser pulses

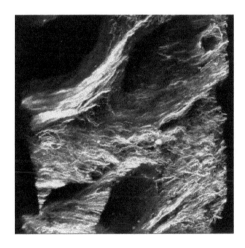

FIGURE 8.8 Collagen fibrils surrounding a mucus gland in human endometrium (see Cox et al., 2003).

(Cox & Kable, 2006). The signal is easily detected in both cryostat and paraffin sections, so routine histological techniques are sufficient for preparing specimens.

Figure 8.8 shows a dramatic example of how SHG imaging can reveal collagen. The tissue is highly autofluorescent, but because the second harmonic signal is both strictly defined in wavelength and propagated strongly forward, the fluorescence does not interfere with the imaging of collagen. The second harmonic signal can be distinguished from fluorescent probes by *wavelength*, *lifetime*, and *direction* of propagation, which means that in essence it can be imaged independently of even multiple fluorescent probes.

Myosin also gives a second harmonic signal (Mohler at al., 2003) and this is beginning to be taken up as a useful tool in the study of muscles and muscle disease. In the plant kingdom SHG has proved valuable in imaging starch and other polysaccharides (Cox et al., 2005).

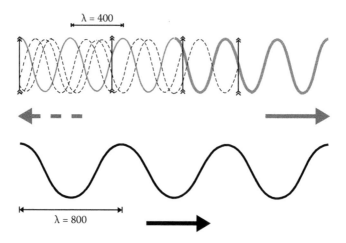

FIGURE 8.9 Randomly spaced fibers propagate the second harmonic in a forward direction. The lower wave is the excitation beam at 800 nm, propagating in the direction of the bottom arrow. The upper row shows second harmonic radiation generated by the randomly spaced dipoles indicated by vertical arrows. Light propagated in the forward direction is in step with both the exciting radiation and light from other dipoles, whereas in the reverse direction radiation from each dipole has a random phase relation with any other, so it tends to cancel out and little propagation takes place.

The second harmonic signal shows a strong tendency to propagate forward. The extent of this forward propagation depends to some extent on the crystallinity of the sample and on how much it scatters light, but with type I collagen in tendon, between 80% and 90% of the SHG signal is propagated forward. Similar proportions apply to starch, but more amorphous forms of collagen give both a weaker signal and a more isotropic one.

The reason for the directionality lies in the coherent nature of the second harmonic generation process. The harmonic is always in phase with the exciting beam. (Each maximum and minimum of the incident waveform translates to a maximum in the harmonic.) Although an individual molecule will radiate the harmonic in all directions, as soon as we have several molecules within the excitation volume, the phase relationship among the individual radiators becomes important. As Figure 8.9 shows, an array oriented along the direction of the beam propagates strongly forward; regardless of the actual spacing (and even if it is completely irregular), all dipoles have the same phase relationship to the exciting beam in the forward direction but are randomly out of phase in the reverse direction. Hence, a microscope that will be used for second harmonic imaging needs a transmitted light detector.

On a practical basis, the coatings used in interference filters tend to absorb strongly at shorter wavelengths. We normally want a narrow-band filter to pass the harmonic signal, and this absorption therefore becomes significant: the narrower the band of the filter, the more layers it contains. Figure 8.10 shows how substantial this effect can become as we approach 400 nm, though some newer filters now perform better. Almost all infrared-blocking filters used to block the excitation beam also cut

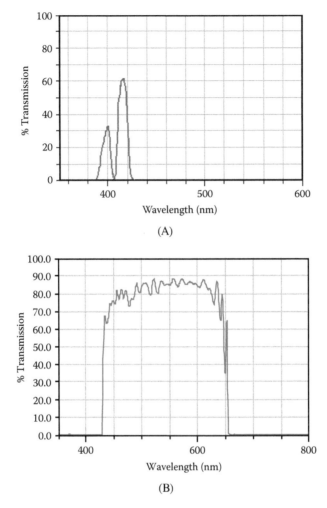

FIGURE 8.10 (A) Curves of two narrow-band filters suitable for detecting SHG signals. Because of the absorbance of the coating materials, at 405 nm (violet, 405/10) only about 35% transmission is attainable, so it is preferable to tune the laser to 830 nm and use the 415 nm filter (blue, 415/10), which passes more than 60%. (B) Curve of the popular 650 sp multiphoton barrier filter. Note that it also blocks anything shorter than 430 nm, so it will not be a good choice for SHG imaging. (Plots courtesy of Chroma Technology Corp. Reproduced with permission.)

off well before 400 nm. The common E650 SP, even though it looks like clear glass, does not pass anything shorter than 430 nm, which will block the second harmonic from any wavelength shorter than 860 nm. When you are doing second harmonic microscopy, therefore, it is a good idea to tune the laser to 830 nm or longer and to check the characteristics of the blocking filter. Often you may need to use a less efficient, colored glass filter rather than an interference filter to block the fundamental wavelength to avoid unwanted short-wavelength cutoff.

SUMMARY

With a number of detectors, a nonlinear microscope can detect harmonic signals along with multiphoton fluorescence, giving a wider range of imaging possibilities than conventional confocal microscopes. Multiphoton microscopy also works well in conjunction with spectral detection, because the entire visible spectrum can be scanned without interference from the excitation wavelength. With high image quality, coupled with better penetration and less damage to living cells than single-photon microscopy, nonlinear microscopy is now regarded as an essential tool for cell biology.

REFERENCES

Campagnola, P.J., Clark, H.A., Mohler, W.A., Lewis, A., and Loew, L.M. 2001. Second harmonic imaging microscopy of living cells. *Journal of Biomedical Optics*, 6, 277–286.

Cox, G.C, and Kable, E.P. 2006. Second harmonic imaging of collagen. In *Cell Imaging Techniques: Methods and Protocols* (Methods in Molecular Biology, vol. 319), D.J. Taatjes and B.T. Mossman, eds. Totowa, NJ: Humana Press, pp. 15–35.

Cox, G.G., Kable, E.P., Jones, A., Fraser, I., Manconi, F., and Gorrell, M.D. 2003. 3-dimensional imaging of collagen using second harmonic generation. *Journal of Structural Biology*, 141, 53–62.

Cox, G.C., Moreno, N., and Feijó, J.M. 2005. Second-harmonic imaging of plant polysaccharides. *Journal of Biomedical Optics*, 10, 024013.

Dedov, V.N., Roufogalis, B.D., and Cox, G.C. 2001. Visualisation of mitochondria in living neurons with single and two-photon fluorescence laser microscopy. *Micron*, 32, 653–660.

Denk, W., Strickler, J.H., and Webb, W.W. 1990. Two-photon laser scanning fluorescence microscopy. *Science*, 246, 73–76.

Dombeck, D.A., Sacconi, L., Blanchard-Desce, M., and Webb, W.W. 2005. Optical recording of fast neuronal membrane potential transients in acute mammalian brain slices by second-harmonic generation microscopy. *Journal of Neurophysiology*, 94, 3628–3636.

Feijó, J.A., and Cox, G.C. 2001. Visualisation of meiotic events in living anthers by means of two-photon microscopy. *Micron*, 32, 679–684.

Gannaway, J.N., and Sheppard, C.J.R. 1978. Second harmonic imaging in the scanning optical microscope. *Optical Quantum Electronics*, 10, 435–439.

Mohler, W., Millard, A.C., and Campagnola, P.J. 2003. Second harmonic generation imaging of endogenous structural proteins. *Methods*, 29, 97–109.

Roth, S., and Freund, I. 1981. Optical second-harmonic scattering in rat-tail tendon. *Biopolymers*, 20, 1271–1290.

Xu, C., and Webb, W.W. 1996. Measurements of two-photon excitation cross-sections of molecular fluorophores with data from 690 to 1050 nm. *Journal of the Optical Society of America B*, 13, 481–491.

Xu, C., Zipfel, W.R., Shear, J., Williams, R., and Webb, W.W. 1996. Multiphoton fluorescence excitation: New spectral windows for biological nonlinear microscopy. *Proceedings of the National Academy of Science USA*, 93, 10763–10768.

Zipfel, W.R., Williams, R.M., Christie R., Nitikin A.Y., Hyman, B.T., and Webb, W.W. 2003. Live tissue intrinsic emission microscopy using multiphoton excited native fluorescence and second harmonic generation. *Proceedings of the National Academy of Science USA*, 100, 7075–7080.

FURTHER READING

Cox, G.C. 2011. Biological applications of second harmonic imaging. *Biophysical Reviews*, 3, 131–141.

Denk, W., Piston, D.W., and Webb, W.W. 2006. Multi-photon molecular excitation in laser-scanning microscopy. In *Handbook of Biological Confocal Microscopy*, 3rd ed., J.B. Pawley, ed. New York: Springer, pp. 535–545.

9 High-Speed Confocal Microscopy

The typical confocal laser scanning microscope (CLSM) has the major disadvantage of being relatively slow in image acquisition. The technological limitation is the need for the fast (X) scan mirror to scan 512 times (or more) per frame, which means that it must oscillate at 2000 Hz in a typical CLSM with a maximum frame rate of 4 per second. Like pie crusts, technological limits are made to be broken, and most manufacturers now make traditional single-spot CLSMs operating at video speed (25 Hz) or faster. This is achieved by using a mirror vibrating at its resonant frequency for the fast scan (Tsien & Backsai, 1995). This enables it to scan much more rapidly but with the penalty that the scan speed cannot be changed. There is a further complication, since something vibrating resonantly moves like a pendulum in a sinusoidal motion rather than linearly. Therefore the speed varies over the width of the frame, being fastest in the middle and slowest at the edges. The manufacturer has to compensate for this or the image will be distorted.

Technological problems are always surmountable, but fundamental physics can be more intractable. As we saw in Chapter 6, even at rather pedestrian frame rates the dwell time on each pixel is the major limitation to obtaining adequate signal-to-noise ratios, so eventually high-speed, single-spot scanners simply run out of signal. This limitation has led to the development of a varied range of microscopes that have as their basic principle the ability to scan more than one point at a time, while retaining the basic confocal geometry.

TANDEM SCANNING (SPINNING DISK) MICROSCOPES

PETRÀN SYSTEM

One such multipoint scanner is, in practical terms, as old as the single-spot configuration. In fact, the first practically useful multipoint scanner was ahead of the single point type, and commercial multipoint and single-point scanners reached market almost simultaneously. The tandem scanning microscope (TSM) uses a spinning disk, or Nipkow wheel, with a spiral pattern of holes in it to scan about 1000 points at any one time (Figure 9.1). As the wheel rotates, the entire specimen is scanned. As one side of the wheel provides the holes for scanning, the opposite half provides the matching pinholes for the detector. (The next scan, of course, takes place half a turn of the wheel later, with the imaging and scanning pinholes swapping places.) The method was devised by Mojimir Petràn and colleagues in Czechoslovakia (Egger & Petràn, 1967; Petràn et al., 1968), and Figure 9.2 shows their design. The original design was manufactured on a small scale by a Czech company. Subsequently,

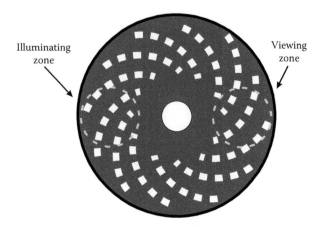

FIGURE 9.1   Diagram of a double-sided Nipkow disk, as used in the TSM. The pinholes are arranged in a multistart spiral, so that many are scanning the field at any one time. At any given time, identical pinholes occupy the illumination and detection areas.

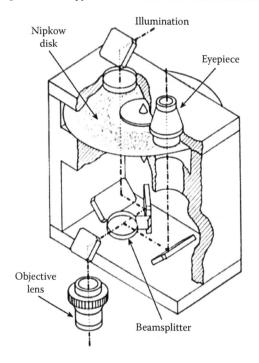

FIGURE 9.2   The original Petràn and Hadravsky TSM design.

U.S. company Tracor-Northern (later Noran) licensed the design and developed it into a production instrument, using mercury or xenon arc illumination, but it is no longer in production.

The great advantage of the tandem scanning system is that it provides real-time operation: The wheel rotates rapidly enough for individual scans to be invisible to the

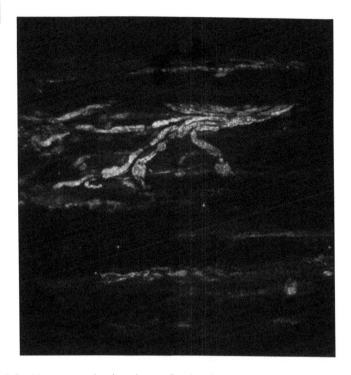

FIGURE 9.3 Neuromuscular junction: reflection image of Golgi silver-stained neurons taken with a Tracor-Northern TSM. (Sample courtesy of Roger Pamphlett.)

eye. Tandem scanning is therefore well suited to studying dynamic processes. Also, a frame typically contains many more scan lines than can be achieved by a single-spot scanning system. The image, a real image produced in space at the plane of the Nipkow disk, is made up of curved scan lines. The image can be viewed through the eyepieces as in a conventional microscope and recorded by a CCD camera or, for greater speed, a video camera. Figure 9.3 shows the image from a Tracor-Northern TSM recorded on film. The circular field of view of the system is apparent. The sample is the neuromuscular end plate; the neurons are stained by Camillo Golgi's classic 19th-century metal impregnation technique (which gives a very effective reflective label for a late 20th-century imaging technology).

In theory, with white-light illumination the image will be in real color, but in practice this is not quite the case. Even the best lenses have some residual axial chromatic aberration, so a range of planes in the specimen is imaged in different colors at the Nipkow disk. This problem is not unique to the TSM, as anyone has discovered who has tried to record a real-color reflection image using red, green, and blue laser lines in a CLSM. But the issue does become more noticeable in the TSM because it is a white-light system and normally used in reflection mode. Nipkow disk systems have an additional problem, because most lenses rely on the eyepiece or tube lens to compensate for lateral chromatic aberration (Chapter 7). The Nipkow disk lies below the tube lens and eyepiece, so off-axis color fringing could also occur. In some cases, the colors seen derive more from optical effects than from the sample; this is certainly the case in Figure 9.3.

The biggest problem with the TSM is that the illuminating source must be spread over a large area with a number of very small holes in it, so most of the illuminating light never reaches the specimen. Typically, the disk stops more than 99% of the incident light, leaving less than 1% to illuminate the sample. Because each illuminating pinhole corresponds to an imaging one, there is no further loss in collecting the image, but of course only a proportion of the incident light is reflected by the specimen. Low light intensity is therefore a major problem with this type of CSM; one needs a very bright light source and a reasonably reflective specimen (or a very high level of fluorescence). In practice, there is often not enough light for fluorescence, and little successful fluorescence work has been done with TSMs; they are mostly used in reflection mode. These microscopes are widely used in the semiconductor industry where their great advantage is that they use a white-light, incoherent source. Laser illumination in the CLSM is highly coherent and causes unwanted effects from interference fringes when you are looking at smooth, highly reflective surfaces, so the TSM gives a more useful image at higher speed.

Inevitably some compromise exists in confocality, because some out-of-focus light returns through the wrong pinhole. The design of the spinning disk determines the number of points being scanned at one time; more spots mean more efficient illumination but poorer optical sectioning. The extent of the problem depends on the thickness of the specimen. With relatively thin samples, no layers are sufficiently far from focus for their out-of-focus disks to reach the next pinhole, but once this threshold is passed, haze and blur start to intrude. The Petràn TSM has the further problem that the two halves of the disk must be identical. This requires very precise construction, but there will always be a limit to how much precision can be achieved. In practice, therefore, the pinholes must be made larger than the Airy disk to allow for inevitable tiny errors in manufacture and alignment, further sacrificing confocality.

ONE-SIDED TANDEM SCANNING MICROSCOPES (OTSMS)

The tricky problem of making both sides of the Nipkow disk identical can be avoided by using the same set of pinholes for both illumination and detection. The light passes directly back through the disk so that the beamsplitter is above the disk rather than below it (Boyde et al., 1990; Xiao et al., 1988). This does introduce another problem, reflection of light from the disk, but this is more of an issue in reflected light for chip inspection (where these microscopes are most used) than in fluorescence.

Current commercial OTSMs targeted at the biological sciences, such as the Becton-Dickinson Carv, use arc lamp or metal halide light sources. Commonly rather large pinholes are used to get a better transmission for fluorescence, with a corresponding reduction in confocality.

MICROLENS ARRAY: THE YOKOGAWA SYSTEM

An elegant (and technologically challenging) solution to the problems of the Nipkow disk microscope was developed by the Japanese company Yokogawa. This system

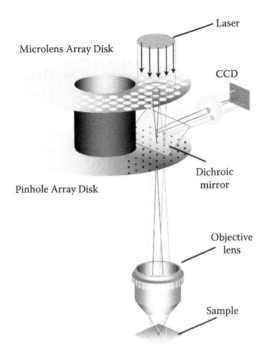

FIGURE 9.4 Diagram of the Yokogawa scanning system, the basis of the Perkin-Elmer and several other real-time confocal microscopes. (Courtesy of Perkin-Elmer, Inc.)

solves the poor light budget of a Nipkow disk microscope with a second, upper disk carrying an array of microlenses each exactly matching one pinhole and thus focusing light on to it. This design increases the light transmission from less than 2% to over 60%. The dichroic mirror is placed between the two disks, so that the returning light passes only through the pinholes and does not return through the microlenses. Therefore, in addition to the improvement in illumination efficiency given by the microlenses, there is the further benefit that reflected stray excitation light comes only from the upper, microlens disk and therefore cannot enter the imaging system. This optical arrangement is illustrated schematically in Figure 9.4.

The Yokogawa system's huge improvement in light budget over conventional TSM and OTSM designs makes possible the real-time imaging of fluorescently labeled living cells. About 2000 points are illuminated at one time, so the signal-to-noise ratio is potentially much superior to a CLSM at equivalent frame speeds. In turn, this means that point intensity can be decreased, thereby reducing bleaching. Figure 9.5 illustrates what can be achieved in a video sequence showing a calcium wave in a cultured cardiac myocyte.

Current Yokogawa modules offer disk speeds of either 300 or 600 rpm, so that very high-speed capture is possible. The image can be viewed directly in an eyepiece but is normally captured by a CCD camera and stored in a computer. Camera sensitivity becomes critical at high frame rates, and intensified or electron-multiplied cameras are often used. (Intensified cameras use an electronic "channel plate"

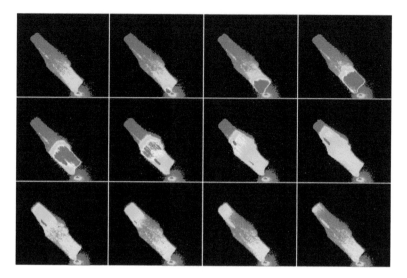

FIGURE 9.5 Calcium wave in cultured cardiac myocytes: 12 consecutive frames from a sequence of 330 taken with the Perkin-Elmer UltraVIEW. (Courtesy of Perkin-Elmer, Inc.)

photomultiplier before the camera, whereas electron multiplication involves a cascade amplification stage at the readout of the CCD. See Chapter 4). Both improve signal, though not necessarily signal-to-noise ratio, and are therefore most useful in cases where otherwise there would be no detectable signal.

Several manufacturers market microscopes based on Yokogawa modules; the best known of these is probably the Perkin-Elmer UltraVIEW. Most of these microscopes use laser illumination with a beam expander to illuminate the array of pinholes. There can be no doubt that this represents the most successful application of spinning-disk technology to the requirements of the cell biologist. However, certain fundamental limits still remain. There will still be crosstalk between pinholes and therefore reduced confocality, particularly with very thick specimens. In a thin sample optical sectioning will be a little different from a point scanner, but the thicker the sample the more crosstalk will reach adjacent pinholes. The area scanned on the sample is fixed, so we cannot zoom in or zoom out by changing the scanned area as we can with a CLSM. Likewise, patterned illumination for photobleaching or photoactivation (Chapter 14) is difficult to achieve unless a separate laser illumination system is introduced. Spinning-disk confocal and CLSM are complementary technologies, not alternatives.

SLIT-SCANNING MICROSCOPES

A different approach to introducing parallelism into confocal microscopes in the interests of higher speed is to abandon the fast X scan. If we illuminate with a line of light, we need to only scan it in the Y direction and can achieve video-rate imaging with moderate mirror speeds (Figure 9.6). The idea of slit scanning goes back a long way; it was first used in a Nipkow disk system to improve the light budget

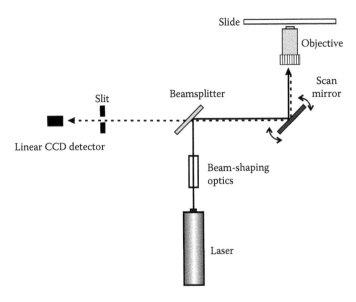

FIGURE 9.6 Diagram of a slit-scanning system. Laser light is formed into a line with a cylindrical lens system and then passes through a beamsplitter. This can be a conventional dichroic but the Zeiss system uses a clever line mirror. The single scanning mirror sweeps the line across the specimen. The returning light is reflected by the dichroic and passes through the confocal slit to a linear CCD detector.

in reflection (Egger et al., 1969). However, modern commercial systems aim rather to provide equivalent imaging modes to conventional laser CSMs, and this is best achieved by conventional linear scanning (Brakenhoff & Visscher, 1992, 1993).

Slit scanning, like Nipkow disk scanning, sacrifices confocality to some extent in the interest of speed, but slit scanning's actual trade-off is different. Slit scanning does not depend on sample thickness in the way that a disk system does, which means that a disk scanner does better with thin samples and a line scanner (potentially, at least) with thick ones. Slit scanning is also anisotropic—resolution, to some extent, is different in different directions because resolution in the Y direction is provided by confocal scanning, but resolution in the X direction is given by conventional widefield optics. However, the major limitation of slit scanning in practice is the technological one of achieving a sufficiently narrow line of light and a sufficiently narrow detection slit and aligning them with each other. Focusing light to a diffraction-limited line (of uniform intensity) is not nearly as simple as focusing it to a diffraction-limited spot.

The previous generation of slit scanners used a mechanical slit, which could be opened or shut to adjust the degree of optical sectioning. However, it is very difficult to make the edges sufficiently uniform, and a trace of dust can make matters even worse. As a result, closing the slit to an optically "optimal" value would produce streaks in the image, as some parts of the slit became much narrower than others. Alignment of a conventional pinhole requires shifting only in X and Y, but to align a slit also requires rotation, and this can be a tricky operation.

The Zeiss LSM5 Live and Duo slit scanners use preset slits deposited as mirror coatings on glass, greatly improving the precision. These slit scanners offer a selection of different slit widths on the same substrate to adjust the degree of confocality. This design has another ingenious feature, which does away with the need for a dichroic mirror. To understand how this trick works, we must think about how the scanning line appears at the back focal plane. The light scanning the sample is a wedge of light, converging on the plane of focus at an angle determined by the NA of the objective. This means that the rays are parallel along the line of the slit but have all possible angles in a direction normal to the slit. At the back focal plane, we therefore see a line as well but a vertical one, reflecting the full range of angles but in a single direction. The cunning trick therefore is to put a mirror with just a vertical line of silvering on it at a plane conjugate with the back focal plane. This reflects the scanning line but stops very little of the returning fluorescence, which comes into the lens in random directions.

Two different approaches have been used for detection of the image in slit scanners:

- The modern solution, as illustrated in Figure 9.6, is to use a linear array CCD detector—Zeiss uses one with 512 elements—which is quick to read out, compared to a two-dimensional CCD array. This approach gives an image on a monitor, as with a CLSM. Unlike Nipkow disk systems, there is no live confocal image that can be seen by the eye.
- The older approach rescans the slit image, either by the reverse side of the same mirror or by a separate mirror, so that a real image is formed in space. This image can then be viewed through a conventional eyepiece, as in the Nipkow disk microscope.

The light budget of a slit scanner is efficient, giving it a strong advantage over the conventional TSM, but the microlens approach to TSM has more or less leveled the playing field here. Slit scanning has the other benefit that the image consists of a straight horizontal line, scanned vertically, which can be transferred directly to a linear CCD array with no aliasing problems. The slit scanner's main disadvantage is that in reality beam and slit widths are substantially larger than the theoretical optimum. As with disk systems, zooming in by reducing the scan area is not easily done, but Zeiss provides a lens system that optically changes the size of the scanned area on the sample to give a limited zoom capability. A slit can never be as efficient at excluding scattered light as a pinhole, so for thin specimens it will be less confocal than a microlens disk system. However the slit suffers no degradation in performance with depth, so it will perform equally well with thick samples as with thin.

MULTIPOINT-ARRAY SCANNERS

It is possible to scan with multiple points without using a spinning disk. If we form a number of beams in a grid array, we can scan that grid across the sample and capture an image more rapidly than with one point. The key is to design a clever grid pattern so that just a single scan motion covers the entire field, yet the pinholes are as

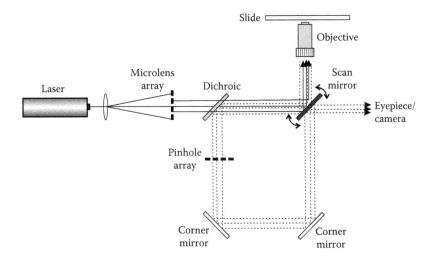

FIGURE 9.7 Diagram of a multifocal scanner (based on the Visitech vtInfinity).

far from each other as possible. Optically, such a scheme is exactly equivalent to a Nipkow disk and has all of the Nipkow disk system's advantages and disadvantages. Because we are scanning with a mirror, the system is mechanically similar to a slit scanner, as will be clear from Figure 9.7 (which is based on the layout of the Visitech vtInfinity microscope). A microlens array forms the multiple beams, and the returning fluorescence passes through a pinhole array, which must be an exact match to the microlens array. The beams are then descanned by the back of the scan mirror so that a real-time live image is available and can be viewed by eye or captured by your choice of camera. Because the scan lines are straight not curved and the scanned area is rectangular, it should be a better match for a camera than the curved scan lines and circular field of view of a disk scanner.

Nikon's Livescan SFC (swept-field confocal) is a similar system, which offers the added refinement of selectable grids with differently sized pinholes. This system also offers the option of using an array of slits for maximum intensity and reduced confocality. (Here, it is almost moving into the territory of structured illumination systems, which are discussed below.) Nikon also uses a separate mirror for rescanning rather than using the rear of the same mirror.

Multiphoton excitation has much to offer in multipoint scanning. Because no pinhole is needed for detection, the problem of aligning the source and detection pinholes is absent. We can therefore expect better confocality and resolution, albeit offset by the longer wavelength. The disadvantage is that, since that two-photon excitation is nonlinear, we cannot split the light too many ways, or we will not get effective excitation. The LaVision Trimscope is a multipoint scanner that operates only in multiphoton mode. The beam from the laser is split into up to 64 beamlets, using a 50/50 beamsplitter and mirrors (Figure 9.8). Each beamlet is identical in power, but each has traveled a slightly different path length. This design also eliminates crosstalk; scattered light from one beam does not affect the excitation of fluorescence by

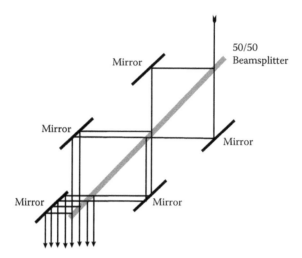

FIGURE 9.8   Beam multiplexing with a 50/50 beamsplitter. A single input beam is split into eight output beams of equal intensity but unequal path lengths.

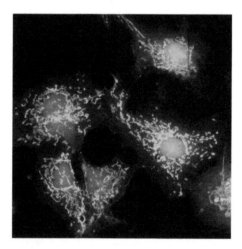

FIGURE 9.9   Vital-stained mitochondria in HeLa cells, imaged with the LaVision Trimscope. (Courtesy of LaVision Biotec Gmbh.)

its neighbor since it arrives too early or too late (Fittinghoff & Squier, 2000; Nielsen et al., 2001). Unlike any other multipoint system, the Trimscope is fully confocal. The image is collected by a CCD camera (Figure 9.9). With a maximum of only 64 points (the number is user-selectable, by factors of 2, naturally), we cannot collect a full-frame scan with only Y scanning, so the speed improvement may be less than that offered by the other systems described so far in this chapter. The choice is between capturing 64-pixel-wide strip images at a very high speed or capturing full-frame images more slowly. But it is still 64 times as fast as a single-point scanner.

STRUCTURED ILLUMINATION

The topic of structured illumination appears here because it is a logical sequel to slit and multipoint scanning from the technical perspective, although in most current commercial applications it is not a high-speed technique. Throughout this chapter the general principle has emerged that just about any obstruction in the image plane tends to remove out-of-focus light more than in-focus light. If we were to take a picture through a horizontal pattern of bars obscuring half the field, the bars would block some proportion of out-of-focus light, so that the 50% of the image that we could see would contain less out-of-focus blur than a normal widefield image. If we then shifted the bars to take the other half of the field, we could combine the two images, strip by strip, and get a final image in which all in-focus information is present, but the intensity of out-of-focus layers is substantially reduced.

Structured illumination imaging systems take this principle one stage further. By shifting the area covered progressively (rather than just on or off), we can observe which structures are changing with each movement (which means they are in focus) and which are not (which means that they are out of focus). Simple image processing (essentially little more than subtraction) can then further reduce the out-of-focus contribution. This approach is the basis of commercial structured illumination systems, such as Optigrid and the Zeiss Apotome.

The grid is moved mechanically, which means that the microscope needs no modifications beyond a slider containing the grid mechanism. Three exposures are taken per frame; typically, these take about one second, so it is not a high-speed technique. (In principle, however, it is easy to think of ways in which the technique could be made much faster.) Structured illumination currently provides a low-cost route to optical sectioning. If you simply require single images free of out-of-focus haze, rather than full three-dimensional imaging, structured illumination is a very effective solution.

REFERENCES

Boyde, A., Xiao, G.Q., Corle, T., Watson, T.F., and Kino, G.S. 1990. An evaluation of unilateral TSM for biological applications. *Scanning,* 12, 273–279.

Brakenhoff, G.J., and Visscher, K. 1992. Confocal imaging with bilateral scanning and array detectors. *Journal of Microscopy*, 165, 139–146.

Brakenhoff, G.J., and Visscher, K. 1993. Imaging modes for bilateral confocal scanning microscopy. *Journal of Microscopy*, 171, 17–26.

Egger, M.D., and Petràn, M. 1967. New reflected light microscope for viewing unstained brain and ganglion cells. *Science*, 157, 305–307.

Egger, M.D., Gezari, P., Davidovits, P., Hadravsky, M., and Petràn, M. 1969. Observation of nerve fibers in incident light. *Experientia* (Basel), 25, 1226–1226.

Fittinghoff, D.N., Wiseman, P.W., and Squier, J.A. 2000. Widefield multiphoton and temporally decorrelated multifocal multi-photon microscopy. *Optics Express*, 7, 273–279.

Nielsen, T., Fricke, M., Hellwig, D., and Anderson, P. 2001. High-efficiency beamsplitter for multifocal multiphoton microscopy. *Journal of Microscopy*, 201, 368–376.

Petràn, M., Hadravsky, M., Egger, M.D., and Galambos, R. 1968. Tandem scanning reflected light microscope. *Journal of the Optical Society of America*, 58, 661–664.

Tsien, R.Y., and Backsai, B.J. 1995. Video-rate confocal microscopy. In *Handbook of Biological Confocal Microscopy,* 2nd ed., J.B. Pawley, ed. New York: Plenum Press, pp. 459–478,

Xiao, G.Q., Corle, T.R., and Kino, G.S. 1988. A real-time confocal scanning optical microscope. *Applied Physics Letters*, 53, 716–718.

FURTHER READING

Pawley, J.B., ed. 2006. *Handbook of Biological Confocal Microscopy*, 3rd ed. New York: Springer.

10 Deconvolution and Image Processing

DECONVOLUTION

When we look at a single, bright point in a conventional widefield microscope, the image of the light is spread out into a complex three-dimensional shape, which resembles a couple of vases placed base to base or the old-fashioned children's toy called a diabolo (Figure 10.1). At the focal plane (the base of these vases or the center of the diabolo), we see a cross-section of this shape, which is the familiar Airy disk. The overall shape is known as the *point spread function* (PSF), that is, the function describing the spread of light emanating from a point. It is a much more complex shape than the confocal PSF (Chapter 5).

So much for the image of a point. The interesting question is to what extent can we consider the image of a whole, complex specimen as the sum of the PSFs of all the points within it (Figure 10.2). Clearly, if we can determine this, then we can predict exactly the image we are going to get from a particular specimen (provided that we know the PSF). The answer is that we can indeed consider the image-forming process in this way, provided it meets two conditions: It must be linear and shift invariant.

> *Linear* means that the image of the whole object is the sum of the images of its parts. For this to be true, the image of a point must not be influenced in any way by neighboring points. An image is not always linear: In the case of differential interference contrast (DIC), for example, the contrast depends on the difference in refractive index between a point and its neighbor (Chapter 2). In fact, if one thinks about the basis of Abbe imaging theory, which treats the specimen as a series of diffraction gratings, it is clear that the linear model will never be true for a conventional brightfield image. It will, however, be true to a reasonable approximation for a fluorescence image, because the fluorescence from each point is incoherent: it cannot interfere with its neighbor.
>
> *Shift invariant* means that the image-forming process is the same everywhere: The PSF is the same for a point on the edge of the image as for one in the center. Any lens has some off-axis aberrations, so this will never be perfectly true. However, a digital camera usually captures only the center of the field, so with a good lens we can take shift invariance as applying to a first approximation. We will, however, encounter problems going down into a sample with a dry lens, because spherical aberration distorts the PSF as we go beyond the point of correction.

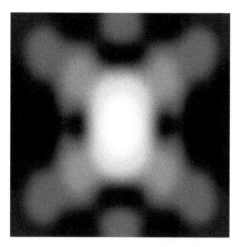

FIGURE 10.1   Vertical section through the point spread function (PSF) of a widefield microscope. The intensities are plotted on a logarithmic scale to make the outer part visible. (Courtesy of Colin Sheppard.)

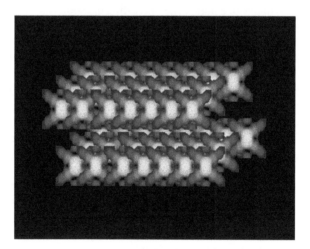

FIGURE 10.2   Two image planes as collections of PSFs.

Having established these two prerequisites, we can discover the PSF by forming the image of a point of light (a tiny fluorescent bead will do in practice) and measuring where the light ends up (for example, by focusing a CCD camera through a series of closely spaced planes, capturing an image each time). We can represent this PSF by a series of numbers, each denoting an intensity, set out in a 3D array. If we now likewise consider our specimen as a series of points in a 3D array, we can calculate the image our microscope would give by going through this array, point by point. At each point, we take the intensity of this particular spot and distribute this light

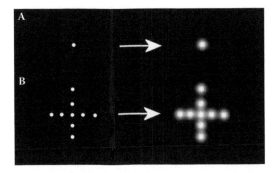

FIGURE 10.3   (A) A single point is imaged as the PSF by a microscope. (B) In a fluorescence image we can go through the whole specimen convolving each point with the PSF to generate the image.

around the matching point in our image array according to the values given by the PSF array (Figure 10.3). We then move to the next spot and do the same thing, then the next, and the next, and so on.

Mathematically, this is known as a convolution. We *convolve* one array of values (our object) with another (the PSF) to give a third: the image. Another function that often crops up in this field is the *optical transfer function* (OTF), which is simply the Fourier transform (FT) of the PSF. Mathematically, it is often easier to handle convolutions in reciprocal (Fourier) space, because the FT of the convolution of two functions is simply the product of the FTs of the individual functions; multiplication is less tedious than convolution.

In principle, if we can compute our image by convolution, surely we can reverse the calculation and turn our blurred, real-life image into a perfect representation of our original object? That transformation is the goal of *deconvolution*. One might propose to take a Fourier transform of our image, divide it by the OTF, and thereby get the FT of the original object. In practice, the process is not so simple because the OTF goes to zero in some places. We can multiply by zero to get the image from the object, but we cannot divide by zero to get the object back from the image. If we cheat by treating the null values as very small numbers rather than zero, the result hugely multiplies the contrast in directions in which no real information was transmitted, and our image is at risk of being drowned by random noise.

The aim of deconvolution algorithms is, therefore, to bring computational ingenuity to bear and get an effective reversal of the convolution that formed our image. Deconvolution involves some assumptions: that the image contains no negative values, that transitions are reasonably smooth, and so on. Deconvolution algorithms vary widely in efficiency and, in parallel with this, in the time they take to run.

One of the oldest, simplest, and least effective techniques is the *nearest-neighbor method*, which takes each slice in turn and considers the likely contribution from the slices immediately above and below (treating them as if they were approximations to the "real" distribution of light). This out-of-focus light is then subtracted from the slice being analyzed. The result removes some haze but worsens the signal-to-noise

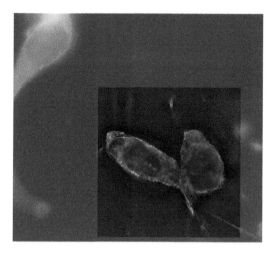

FIGURE 10.4   Cultured HeLa cells. The large field shows the original image stack and the window shows a subset deconvolved with an inverse filter.

ratio. Nearest-neighbor cannot be considered an effective method, though it may give us a preview of whether our data set is worth deconvolving with a more rigorous and time-consuming method.

Inverse filtering is a single-pass process that attempts to divide the FT of the image with the OTF, making some approximations to overcome the zero problem (Figure 10.4). Inverse filtering generally does much better than nearest-neighbor, but some degree of filtering to avoid generating spurious structures from random noise will probably be needed.

More accurate methods go through a series of iterations, gradually converging on to a "best" or most consistent value (Figure 10.5). The aim of these methods is to get a result that when convolved with the PSF does give the observed image. They differ in the type of constraints applied and in how they obtain and use the PSF data.

Some methods use a calculated PSF on the basis that this will be free of the random noise inevitably associated with an experimentally measured PSF. Others use a measured PSF on the equally rational basis that it will show the actual performance of the particular lens being used (Carrington et al., 1995). *Blind deconvolution* (Holmes, 1992) makes only general assumptions about the form of the PSF, and then refines this guess in successive approximations (taking therefore even longer to compute) to find the actual PSF that gives the "best" result.

Defining the best result—in other words, assessing whether the algorithm is converging or diverging—is an essential part of the deconvolution process. One approach is to use a *maximum likelihood estimation* (MLE; Holmes, 1988). Assuming that the imaging process is essentially random, in other words, that it is a statistical process, one calculates the likelihood of the computed object giving rise to the observed image. The best result is the one that has the highest likelihood value. An alternative approach is to use entropy: The "real" structure is likely to be the one with the highest degree of order, in other words, the lowest entropy.

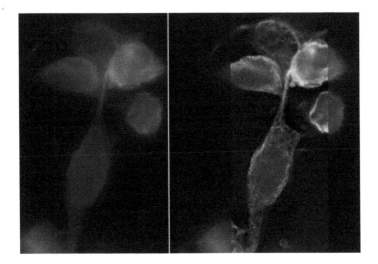

FIGURE 10.5 Constrained iterative deconvolution of the same HeLa cells seen in Figure 10.4. (Left) Original image; (right) deconvolved.

The advantages of deconvolution are:

- *Greater light-collection efficiency.* Because all points on the image are sampled at once, a 1-second, 512×512 pixel frame captured with a CCD camera is equivalent to scanning the same size frame at the same light intensity for 262,144 seconds in a confocal microscope. In practice, because it uses a laser, the confocal uses a much higher light intensity, but even so the CCD has a great advantage. It can capture pictures more quickly with a better signal-to-noise ratio. This speed advantage does not extend to seeing the final images. The computation, even with fast computers and signal processing boards, takes much longer than even the slowest confocal microscope.
- *The ability to use any light source to excite fluorescence.* Lasers have only fixed wavelengths, but modern widefield light sources offer pretty much any wavelength the objective can transmit (Chapter 3). This ability is particularly useful in excitation ratio imaging (Chapter 14), which in most cases is not possible with laser illumination.

The disadvantages of deconvolution are:

- *The algorithms do not cope equally well with all types of objects.* Some structures in the reconstructed images are imaged more faithfully than others. Unless the original sample is fairly simple and predictable (e.g., fluorescent in situ hybridization (FISH) staining, spindle microtubules), it can be difficult to know how much trust to put in the reconstruction.
- *The computation is slow.* The advent of ever-faster computers has helped to some extent, but designers have tended to take advantage of that to make algorithms more complex and sophisticated.

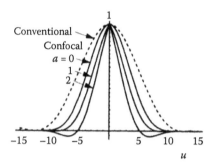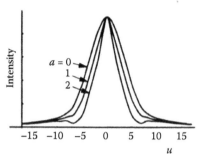

FIGURE 10.6 Effects of a 1D single-pass deconvolution on the vertical profile of the PSF of (left) a point object and (right) a planar object. The factor a is a scaling factor depending on the degree of under- or oversampling, 0 meaning no deconvolution. The effect is to give a clear minimum to the function and hence improve the effective resolution in Z. (See Cox & Sheppard, 1993.)

DECONVOLVING CONFOCAL IMAGES

Confocal microscopes also convolve the object data with a PSF, and deconvolution can therefore also be applied to confocal images. Because the confocal PSF is much less complex than the widefield PSF, approximating an ellipsoid (Figure 5.6), this is a simpler process than deconvolving widefield images. On the downside, however, confocal images typically have a poorer signal-to-noise ratio than widefield, limiting the gains that can be achieved. One particularly useful approach with confocal images is deconvolving just in the Z direction where the confocal performance is worst (Cox & Sheppard, 1993).

Figure 10.6 shows the axial response function (i.e., the profile in Z of the PSF) for a confocal microscope on small and extended objects. The response is broad, as we know, but it also suffers from the lack of any clear minimum, especially on extended objects. A simple one-pass deconvolution operation not only narrows the response, it also provides a clear minimum. The latter point is particularly important, because our eyes like to find edges, so this algorithm can greatly improve the appearance of three-dimensional reconstructions from confocal data sets. The effect on biological images is striking, largely because the shape of the confocal spread function, particularly for extended objects, gives a visual impression of worse depth resolution than is actually achieved. Modifying this response gives a modest improvement in real resolution and a big improvement in the perceived appearance of the image (Figure 10.7). This aspect is discussed further in Chapter 11, in the context of 3D rendering, and is illustrated in Figure 11.15.

IMAGE PROCESSING

The overall aim of this book is to encourage and educate the reader to take the best possible pictures. But there are times when sample or equipment limitations mean

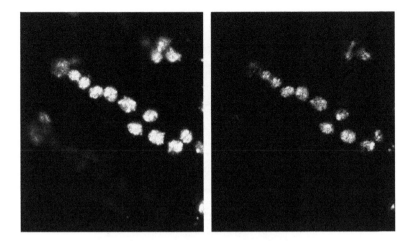

FIGURE 10.7 The effect of 1D deconvolution on a confocal image stack of chloroplasts in a *Selaginella* leaf. (Left) Original image; (right) deconvolved. One section from the stack is shown, and the out-of-focus parts of the original image are clearly removed by the processing. (See Cox & Sheppard, 1993.)

that the best possible picture still has deficiencies, and that is where image processing comes in. Image processing can never restore something that is not there, but it can be very effective in making what *is* there visible. Many of the operations described in this section are available within confocal or widefield acquisition software, but specialist software offers more. Appendix D presents a full tutorial on practical image processing with the public-domain software ImageJ.

GRAYSCALE OPERATIONS

The values that make up our digital image are numbers, and we can manipulate them like any other numbers. If we add a fixed value to each one, we make the image brighter. If we subtract a fixed amount from each one we make it dimmer. We can also do this to remove a gray background from the image and make it look a more presentable black. If we multiply all the pixels by a fixed amount, we increase the contrast, boosting the bright pixels more than the dim ones. Likewise, dividing all the pixels by a constant reduces the contrast. By combining these operations, we can do one of the most common image-processing actions: normalizing the contrast (also called *histogram stretching*). The idea is to make the darkest pixel 0 and the brightest 255. Here is what happens when we click the "Stretch Histogram" button:

1. The image is scanned and the value of the darkest pixel is found.
2. That value is subtracted from all pixels, so our darkest shade is now black.
3. The image is scanned and the value of the brightest pixel is found (or maybe the software was smart enough to do that in the first pass, and then subtract).
4. The value of every pixel is multiplied by the amount needed to make the brightest one equal 255.

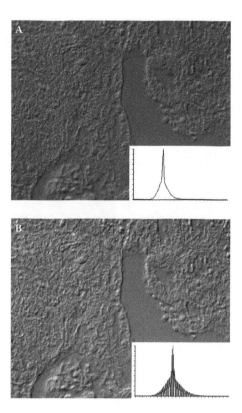

FIGURE 10.8 DIC image of a kidney section. (A) Original image, dark and lacking in contrast, the histogram shows gray values only using a part of the available contrast range. (B) After contrast stretching, a much more pleasing image. The histogram shows a broader spread of values but with gaps in it.

Figure 10.8 shows this in practice. A rather dim DIC picture has a histogram with the gray values close together and in the low part of the range. Stretching the contrast gives a much more usable picture. Because we cannot have more gray values than we started with, there are now gaps in the histogram. At this scale the gaps do not cause any visible problems, but if we had very few gray values in the original image they would cause posterizing (Chapter 6).

There are other ways to remap gray values. Another common problem we encounter is caused by the fact that the CCD or confocal image is linear; the numbers represent the number of photons measured at each pixel. However, display and printing devices typically are not linear, so midtones are displayed darker than they should be. (This point is discussed in Chapter 6.) To correct the problem, we use a curve function to map the gray values so that 0 and 255 remain the same but an intensity of 128 is raised to, for example, 150, and intermediate values are adjusted accordingly. This is a very useful function, though, sadly, there is no guarantee that the adjustment that makes an image perfect on the monitor will be correct for a printer or a data projector.

IMAGE ARITHMETIC

We can also do mathematical operations involving two images, where we modify each pixel according to the value of the corresponding pixel in the second image. Figure 10.9 shows a common example of this: subtraction to correct for bleed-through, where one channel appears in the other. In this case, the fluorescein isothiocyanate (FITC), which fluoresces green, has a substantial "tail" of fluorescence in the red region, and therefore appears faintly in the red channel. After scaling the channel 1 image to the level of the bleed-through (Figure 10.9C), we can subtract it from the red channel image and get a picture free of the bleed-through (Figure 10.9D). Another use for subtraction can be to identify moving particles; the static background that has not changed between exposures is removed, but the moving objects remain.

In theory, any sort of arithmetic can be done between two images, but the other operation that is typically useful to cell biologists is taking the ratio of two images. Doing so can give us an internal standardization, independent of dye loading, fading, and lamp or laser fluctuations. Taking this ratio is most useful in such situations as

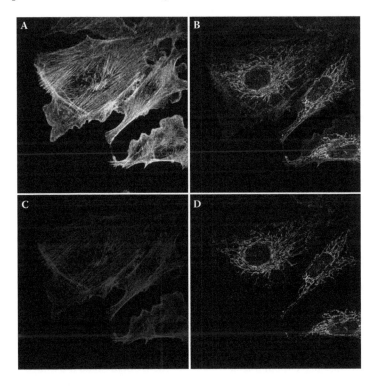

FIGURE 10.9 Image subtraction used to correct for bleed-through. (A) FITC antitubulin labeling of microtubules in a HeLa cell, channel 1. (B) $DiOC_6(3)$ labeling of ER in channel 2. There is considerable bleed-through of the FITC from channel 1. (C) The FITC image a scaled to match the intensity of the bleed-through. (D) Image B minus C, giving us an ER image free from bleed-through.

measuring ion concentrations or membrane potential, and is considered further in Chapter 14. Here, the important issue to bear in mind is the arithmetical point that if we are doing division between two images, we will rapidly run out of numbers if we only work in 8 bits. Twelve-bit depth is pretty much the minimum for any ratiometry.

CONVOLUTION: SMOOTHING AND SHARPENING

We have already seen that the process of image formation can be regarded as a convolution between two sets of numbers. The logical consequence of this is that we can choose to do a convolution ourselves and thereby transform an image in some way or another. In fluorescence microscopy, where photons are always in short supply, the most common task is removing or minimizing noise. This may be just to make an image presentable, but it is pretty much a necessity if we want to do any form of deconvolution or image reconstruction on our data, because noise would otherwise swamp our real data.

Consider a very simple convolution matrix or kernel:

$$
\begin{array}{ccc}
1 & 1 & 1 \\
1 & 1 & 1 \\
1 & 1 & 1
\end{array}
$$

What will this do if we convolve it with an image? The operation is simple, if tedious. Superimpose this grid of numbers on the first set of nine pixels at the top left corner of our image. Let's say that they are the following set. We multiply the nine values in our grid with the corresponding values in our nine image pixels (which in this case leaves them just the same). We add these nine values together (887) and divide by the sum of the values in our convolution matrix (9), which gives us (rounded) 99. This gives us the value for the central pixel of our 3 × 3 matrix in the output image. The convolution operation always works on one pixel at a time, modifying by comparing it with its neighbors.

127	136	56		127	136	56
100	75	90	becomes	100	99	90
101	105	97		101	105	97

Having done this pixel, we move our grid one pixel along on our input image and do the same thing to generate the next pixel of the output. This simple convolution actually replaces each pixel with the mean value of the nine in our grid. You will see, too, that this convolution will not process the outermost pixels: the top and bottom rows, and the left and right columns, in our image. We can deal with those pixels in different ways—leave them unchanged or cut them out to make the image smaller—but the best solution is probably to duplicate these outermost pixels so that we can cover the whole image.

So what did our convolution do? Figure 10.10 shows the result. Figure 10.10A is the noisy image from Figure 6.3A, and Figure 10.10C is the same image after

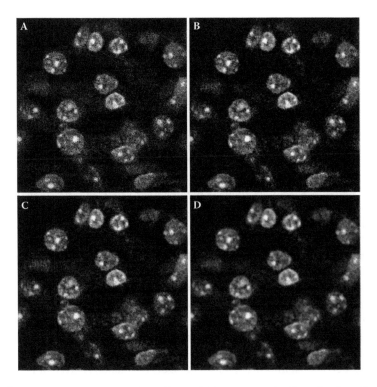

FIGURE 10.10   (A) Noisy image from Figure 6.3A, (B) smoothed by a 3 × 3 median filter, (C) smoothed by a 3 × 3 (mean) smoothing filter, (D) smoothed by a 5 × 5 weighted smoothing filter.

convolving with our simple kernel. The noise has been effectively smoothed out without destroying the image: This kernel is a *smoothing filter*. If we extend the kernel to encompass more pixels, we smooth the image more:

$$
\begin{matrix}
1 & 1 & 1 & 1 & 1 \\
1 & 1 & 1 & 1 & 1 \\
1 & 1 & 2 & 1 & 1 \\
1 & 1 & 1 & 1 & 1 \\
1 & 1 & 1 & 1 & 1 \\
\end{matrix}
$$

This is a 5 × 5 kernel, and to give the original pixel value a bit more weight in the output we have made the center value 2. Figure 10.10D shows the result and the extra smoothing is clear.

An alternative approach to smoothing is to use a *median filter*, which is not actually a convolution, although it works in a similar manner to one. A median filter looks at all the pixels within the region defined by its kernel. The central pixel is then replaced by the median value of these pixels. In general, median filters have the property of effective noise removal, with a relatively low penalty in terms of reduced resolution.

Using our same set of numbers, we see that in the nine numbers of our 3 × 3 matrix, 100 is the median, because there are four values above it and four below. This therefore goes into the output pixel:

127	136	56		127	126	56
100	**75**	90	becomes	100	**100**	90
101	105	97		101	105	97

The effect on our test image is shown in Figure 10.10B. Because sorting numbers is slower in a computer than arithmetical operations, median filters were once thought of as being time consuming, but modern computers are so fast that time is no longer an issue. Like other filters, a median filter can be any size matrix, and the matrix need not be square; a cruciform or circular matrix, where we look only at the pixels that adjoin at sides not corners, is also a popular choice.

The main noise component in confocal fluorescence imaging is photon sampling or shot noise. This, by its nature, is randomly distributed in all three dimensions. Applying noise-reduction filtering only to individual, two-dimensional slices modifies the noise distribution anisotropically and is clearly not the optimal approach; effective noise removal requires filtering in as many dimensions as the data set contains (Cox & Sheppard, 1999). Here, the cruciform matrix is definitely at an advantage; going from two to three dimensions adds just two more voxels (those above and below) to the calculation, whereas a square matrix adds eighteen (nine above and nine below). An example of 3D median filtering applied before deconvolution is given in Figure 11.15.

Smoothing comes always at the expense of resolution. Averaging sufficient images to get a noise-free result is always going to be the better option. However, doing so is not always possible, and if we collect with just a little oversampling (say, three pixels in the minimum resolved distance) we will have enough resolution in hand to apply a mild smoothing without losing resolution.

Smoothing is not the only thing we can do with convolution filters. There are all sorts of tricks we can play, for example, creating embossed or bas-relief effects by making the filter matrix nonsymmetrical. However, the only other operation likely to be useful to us as microscopists is sharpening a blurred image. If we take the following kernel,

$$
\begin{array}{ccc}
 & -1 & \\
-1 & 5 & -1 \\
 & -1 & \\
\end{array}
$$

which is an example of a nonsquare kernel

127	126	87		127	126	87
100	**99**	90	becomes	100	**74**	90
101	105	97		101	105	97

The figures add up to 75, and no division is needed because the kernel sums to 1. (You may have noticed that the set of numbers has been changed. This is because the previous set was very noisy and noisy images do not work well with

FIGURE 10.11   A sharpening filter applied to an out-of-focus photograph.

a sharpening filter. Try the calculation with the figures in the previous examples.) The filter's effect is to make the target pixel more different from its neighbors, and it is this effect—magnifying any difference between neighboring pixels—that does the sharpening.

The effect of this filter on an out-of-focus photograph, shown in Figure 10.11, is striking (it is a pretty severe sharpening filter). Fluorescence images are usually too noisy for sharpening filters to be of much use, and confocal microscopes, by definition, cannot take out-of-focus images, but sharpening filters are quite applicable to brightfield or phase images. One day, a sharpening filter might save the day when the only micrograph you have is out of focus and the specimen has been lost.

REFERENCES

Carrington, W.A., Lynch, R.M., Moore, E.D.W., Isenberg, G., Fogarty, K.E. and Fay, F.S. 1995. Superresolution three-dimensional images of fluorescence in cells with minimal light exposure. *Science*, 268, 1483–1487.

Cox, G.C., and Sheppard, C.J.R. 1993. Effects of image deconvolution on optical sectioning in conventional and confocal microscopes. *Bio-Imaging*, 1, 82–95.

Cox, G.C., and Sheppard, C.J.R. 1999. Appropriate image processing for confocal microscopy. In *Focus on Multidimensional Microscopy*, vol. 2, P.C. Cheng, P. P. Hwang, J.L. Wu, G. Wang, and H. Kim, eds. Singapore: World Scientific Publishing, pp. 42–54.

Holmes, T.J. 1988. Maximum-likelihood image restoration adapted for noncoherent optical imaging. *Journal of the Optical Society of America A*, 5, 666–673.

Holmes, T.J. 1992. Blind deconvolution of quantum-limited incoherent imagery: maximum-likelihood approach. *Journal of the Optical Society of America A*, 9, 1052–1061.

FURTHER READING

Russ, J.C. 2011. *The Image Processing Handbook*, 6th ed. Boca Raton, FL: CRC Press.

11 Three-Dimensional Imaging

Stereoscopy and Reconstruction

Confocal microscopes generate data in three dimensions; the challenge to the user is to maximize the useful information extracted from this data. Running through a stack as a movie can be valuable, but it is difficult to comprehend our sample as a three-dimensional object from such a view. We need to find a way to present the data so that our brain can interpret it as an entity, not as a collection of slices.

We can extract data in two directions automatically as XY and XZ sections. In combination, these views can sometimes give us vital information that we can at least interpret statistically, but visual interpretation of such views is not so easy. Statistical analysis, of course, can be of immense scientific value. But if we cannot get an intuitive view of our data, often we have no idea what statistical questions to ask.

SURFACES: TWO-AND-A-HALF DIMENSIONS

Images of surfaces can be easy to handle. These are sometimes called *2.5D specimens*, because they consist of a two-dimensional entity—the surface—deformed in the third dimension. With 2.5D specimens, we can generate depth-coded maps, and because the information to be displayed is relatively simple, these maps can be extremely effective at displaying the topography.

By taking a maximum brightness projection (see "Simple Projections" section) of such a data set, we can generate a true and accurate extended-focus image showing the surface as it would appear if we had a microscope with infinite depth of field (Figure 11.1). The accuracy is assured because in a confocal data set the image is always brightest at the point of best focus.

In the process of reconstructing such an image, we automatically obtain another form of highly useful information: the depth at which the surface is located. This can also be presented in image format as a depth-coded map (Figure 11.2). Here, brightness represents distance from the objective, so high points are white and low points are black. There were only 22 sections in the data set, so there are only 22 gray values in the image.

Having generated such maps, we can extract a wide range of useful data from them. Figure 11.3 shows how we extract a line profile of depth across the surface, correct it for overall slope, and then use it to obtain standard statistical measures of roughness.

FIGURE 11.1 Extended focus image of a metal surface (a well-worn coin).

FIGURE 11.2 Depth-coded projection of the surface seen in Figure 11.1, generated at the same time.

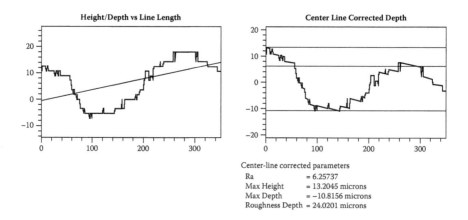

Center-line corrected parameters
Ra   = 6.25737
Max Height   = 13.2045 microns
Max Depth   = −10.8156 microns
Roughness Depth   = 24.0201 microns

FIGURE 11.3 Profile from the depth-coded projection (left); corrected for slope (right) and roughness measurements derived from it.

Many confocal microscopists in materials science look almost exclusively at this type of sample, and the confocal microscope is a powerful tool for their purpose. As cell biologists, though, we are usually dealing with a true three-dimensional data set and need to interpret the whole of our sample as a three-dimensional entity. Doing so is harder than it seems, particularly if the specimen is complex or crowded (and cells are usually both).

PERCEPTION OF THE 3D WORLD

We believe that we see the world in three dimensions. This is not really so; rather we *interpret* the world in three dimensions. The difference in parallax between the views seen by our two eyes (binocular parallax) is one factor in this, often the major factor, but experiments have shown that people can tell that one object is in front of another even when the parallax is below the resolution of the eye.

It is worth listing the extensive range of other factors we use:

- Motion parallax when we move our head or our eyes
- Convergence and focus of our eyes
- Perspective (objects appearing smaller as they become more distant)
- Concealment of one object by another
- Our knowledge of the size and shape of everyday things
- Light and shade

All together, these factors are extremely effective; you need only watch a cricketer or tennis player to realize just how good our spatial perception is. However, take each factor separately in the context of presenting 3D microscope data sets to the human brain and serious problems arise.

Motion Parallax

Many incidents can bring home to us just how important motion parallax is. If we work with electron microscopes, we may have noticed how tilting the sample suddenly makes it seem to leap out at you in three dimensions. Often, one gets the same impression when flying low in a plane that is coming to land: The rapid motion over the countryside suddenly makes it stand up in heightened relief. Fortunately, as you will see later in this chapter, motion parallax is one feature we can call to our aid in presenting optical data sets on the computer.

Convergence and Focus of Our Eyes

When we look at an object that is close to us, our eyes converge, and also muscles in our eyes pull on the crystalline lens to make it thicker and thus shorten its focal length. With relatively close objects this is a vitally important cue, but it is totally lost to us in presenting confocal data. Whatever illusion we are trying to create, the image remains obstinately in one plane, on the sheet of paper or the computer monitor.

PERSPECTIVE

When we collect a 3D optical data set each plane is collected at exactly the same scale. There is no perspective. High-end 3D rendering software can put perspective back into reconstructions, but the simple 3D software included with your confocal system will not.

CONCEALMENT OF ONE OBJECT BY ANOTHER

Our evolutionary history did not design our visual system to deal with totally transparent objects. Unlike insects and birds, we learn enough to cope with windows and glass doors; even so, accidents involving glass are common. In the microscope, when we look at widefield or confocal fluorescence images, or even at a histological section, everything is transparent. Nothing in everyday life prepares us for a world in which we can see through everything. Another of our essential visual cues—objects obscuring their background—is therefore lost. Opacity can be reinstated by software, and as we will see, this is a key factor that differentiates advanced 3D rendering software from the simple utilities supplied with confocal microscopes or as freeware.

OUR KNOWLEDGE OF THE SIZE AND SHAPE OF EVERYDAY THINGS

We have all seen trick pictures that use "wrong"-sized tables, chairs, or whatever to give us a false sense of position. These pictures fool us because the objects are familiar, and we have learned what such objects are like. Although we may know, intellectually, the size of a mitochondrion or nucleus, this knowledge is not hardwired into our visual cortex in the same way. There are no spatial cues in a cell, so we have no option but to do without them.

LIGHT AND SHADE

In productions of Shakespeare's play *Macbeth*, the three witches are always lit from below. However familiar we are with this lighting trick, it always succeeds in making the witches seem weird. Over millions of years, we have gotten used to light coming from above. Self-luminous objects (fireflies, candle flames, burning logs), while not unknown, are not routine in our world. In the world of fluorescence microscopy everything is self-luminous, not lit by an external source. Bringing in directional lighting and shadows (common computer technology nowadays) can be a great help in interpreting complex 3D objects.

LIMITATIONS OF CONFOCAL MICROSCOPY

We face a further problem in creating realistic-looking reconstructions from confocal data because the CSM does not have equal resolution in all three dimensions. We can make the image look better by undersampling in X and Y, using a high numerical aperture (NA) lens at a low magnification, but often we need to capture detail that this technique does not give us. Normally we would expect the Z resolution to be, at best, half as good as the X–Y resolution. We often make matters worse by not taking enough optical sections. To fully exploit the Z resolution, we should take at least 2.3 slices within that figure (the Nyquist criterion), and now that hard disk space is cheap there is no excuse for not doing so, provided our fluorochrome will survive without fading.

We can also improve reconstruction quality by image processing or restoration, which can greatly improve the appearance in the Z axis. This is especially valuable because the shape of the response curve in the Z direction—with no clear minimum—makes the image look considerably worse than it really is (Figure 10.6).

So how can we use our 3D data sets to give us an impression of reality? Other than cutting Styrofoam models (!), there are two approaches: stereoscopy and reconstruction.

STEREOSCOPY

Stereoscopy means producing one view for each eye from our series of slices. In other words, we are providing binocular parallax but nothing else. Stereoscopy has the merit that we are looking in the direction the light goes, so the eye gets the best view of the planes that are best resolved. However, we cannot look around things to determine what is in front of what, and we can get confused by transparent objects.

In a stereo pair, depth is converted into displacement between two images. Distant objects appear at the same place in both pictures; near objects appear displaced toward the center line between the pictures. There is a limit to how much displacement the eye can tolerate; in a real specimen, we converge and focus our eyes differently when looking at near and far objects. In a stereo pair picture, everything is actually at the same plane, so we cannot do this. On a specimen with a lot of depth, therefore, we may have to artificially restrict the displacement—that is, foreshorten the depth—for the sake of comfortable viewing. The resulting stereoscopic image still suffers from the disadvantage that all objects are transparent, and from the lack of motion parallax. If we move our head when looking at a stereo pair, the entire image appears to swing. Our brain interprets this as the image following our head around, because we are trying to look around an object but failing to see what is behind it. Nevertheless a stereo image can be a dramatically effective way to present confocal images, and confocal microscopes usually offer stereoscopic views of confocal data sets within their standard software.

Stereo pairs are generally produced in confocal microscopy by projecting the series of images onto a plane, displacing each plane by one pixel from its predecessor (Figure 11.4) (Chen et al., 1995). Pixel shifting is not geometrically ideal—distant objects are not made smaller, so there is no perspective—but it is not too bad, either. This technique has the disadvantage that the perceived depth is essentially arbitrary, a function of pixel size and section thickness. Figure 11.5 shows the result.

An alternative approach is to rotate the data set to generate the two views (Figure 11.6), in other words, to use two views from a reconstruction (see next section). In real-world stereo, this is a bad thing because it makes parallel lines in one plane converge. In our artificial world, however, distant objects do not get smaller, so the geometry is arguably better than with pixel displacement (Chen et al., 1995).

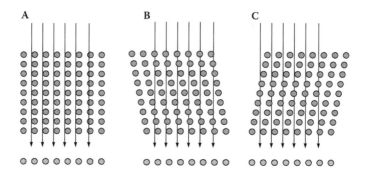

FIGURE 11.4 Pixel shift projection to produce stereo pairs. (A) Direct projection through the stack; (B) projection shifting higher layers progressively to the left, to give the right eye view; (C) the opposite shift to give the left eye view.

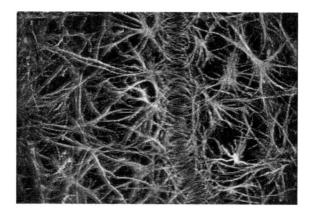

FIGURE 11.5 Red–cyan anaglyph image of astrocytes in the retina. Note how they curve around the blood vessel that runs through the center of the image. It is thought that they actually define a route for the capillary. This anaglyph will work with red–cyan or red–green glasses. (Specimen courtesy of Jonathan Stone.)

At least we have a defined angle between the views for each eye, which is difficult to achieve with pixel displacement. However, in practice there is very little difference between the two algorithms.

The resulting stereo pair can be presented on screen as an *anaglyph,* that is, with one image in red and the other in green or cyan, and viewed through appropriately colored glasses (Figure 11.5). The original images must be monochrome, single channel, and without false-color palettes.

The lack of color is rather limiting, and one can get viewers for looking at stereo pairs placed side by side on the computer screen, so that we can see multichannel or false-color images in stereo. (Side-by-side stereo is easier to do on paper, and journals often publish images side by side for viewing with a folding stereoscope.) A more sophisticated approach requires the viewer to wear special glasses containing liquid-crystal shutters and linked to the computer by an infrared receiver. The left and

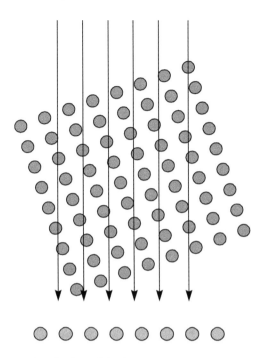

FIGURE 11.6   A rotated projection of the data set.

right images alternate rapidly on the monitor while the glasses switch between the two eyes in synchronism. This system is a catalog option on most Silicon Graphics computer systems and is available as a third-party addition to PCs and Macs, but it is expensive. As an alternative, using conventional projectors you can project the two images through polarizing filters and view them with Polaroid glasses, a favorite technique in IMAX cinemas. It is necessary to use a silver screen; a white one does not preserve polarization.

THREE-DIMENSIONAL RECONSTRUCTION

The most versatile approach to creating realistic 3D impressions is to produce a series of computer-generated views of our specimen, mimicking the views we would get if we looked all around it as a real object. This is computer 3D *reconstruction*, or *volume rendering*: the application of formulas or algorithms that can be implemented on a computer to create three-dimensional projections from data sets. Often several projections are created from different viewpoints, and these can be played in quick succession to form rotations or animations. In this way we can bring in the element of motion parallax, which can enhance our understanding of the data's three-dimensional characteristics.

There are several ways to render volume data sets on a computer, and they can give quite profoundly different views of our image. In general, there is always a trade-off between speed and preservation of accuracy and detail.

In most cases, the data sets we get from our microscope will be treated as a set, or grid, of XYZ coordinates of points in space, with an intensity value for each point. Sometimes, however, we can use a simpler data set containing just the X, Y, and Z coordinates of specific objects of interest. We may therefore need, or prefer, to first convert the former full, data set into the simpler case by deciding which points actually correspond to our "object" then scrapping the rest.

TECHNIQUES THAT REQUIRE IDENTIFICATION OF "OBJECTS"

The oldest technique for rendering volumes is to assemble a stack of *wire-frame* outlines (Figure 11.7). This technique requires minimal computing power and memory, because each section image is represented only by a few lines, but it sacrifices almost all detail of the volume. In the early days of computer reconstruction, it was normal to identify the "objects" by drawing around them on a graphics tablet or with a light

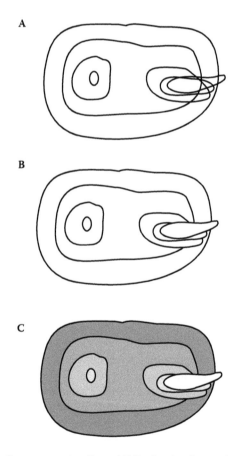

FIGURE 11.7 Wire-frame reconstructions. (A) Basic wire-frame; the "bump" on the left is easy to interpret but the overlapping projection on the right causes confusion; (B) with hidden lines removed, which makes the structure much clearer; (C) shaded with gray levels corresponding to depth, for even greater clarity.

pen on a screen. Later came software that allowed objects to be segmented, that is, recognized by grayscale values, and the software would draw around them automatically. The next level of sophistication removes lines from the frames that are covered by a higher plane of the object (hidden-line removal, Figure 11.7B). It is interesting to recall that when the first commercial confocal microscopes reached the market in 1987, hidden-line removal was still regarded as a challenge for a personal computer. Shading in the frames further enhances the effect (Figure 11.7C), but nevertheless this approach is now just of historical interest.

Surface extraction is in principle a much more advanced development of the wireframe approach. The original confocal data sets consist of a set or grid of XYZ coordinates of points in space with an intensity value for each point. We can reduce this to a simpler case by segmentation to identify the object or objects we wish to render, and then extract a reduced data set containing just the coordinates of the surfaces of our objects. Surface extraction typically operates by looking either for gradient changes or for defined intensity changes to identify the boundary of an object (Figure 11.8).

Crudely representing the surface by individual voxels can make it look very blocky, unless the data set is very oversampled (Figure 11.9). By using geometric rendering—fitting tiny polygons to the points defining the surface of the object—we can produce a more accurate view of the surfaces of the objects we have extracted. One popular approach is the so-called marching cubes algorithm (Cline et al., 1988),

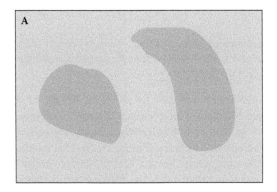

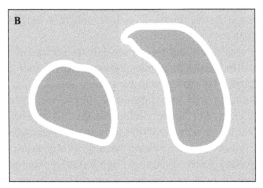

FIGURE 11.8 Identification of the surface of objects by segmentation according to gray values.

FIGURE 11.9   Representing a smooth surface with cubic voxels can look very "blocky."

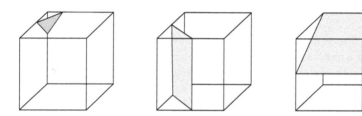

FIGURE 11.10   The marching cubes algorithm. We have a library of 256 different truncated cubes and replace the gray value at each point on the surface by a number indicating which of 256 different shapes best fits that point.

which avoids the blockiness problem by having a library of 256 different ways in which a cube can intersect with a surface. Figure 11.10 shows three examples. When the surface is encountered, the algorithm checks adjacent voxels and picks the appropriate cube value from its library (Figure 11.10). This number (0–255) is stored instead of a grayscale value. Only the details of the surface are stored, greatly reducing the amount of data to manipulate. At the drawing stage—when the image is actually produced on the screen—more pixels are needed to represent the polygons, but this is only a 2D image, so the overhead is small relative to the gain from manipulating a reduced 3D data set.

Geometric rendering is fast, but we can see only the surface of our object. Internal detail is lost and so (in simple implementations) are the actual grayscale values of the surface of the object. By storing a 16-bit number for each point, we can preserve grayscale information, as well as a cube value, improving realism at the expense of doubling the data size. Surface rendering can give an excellent view of a specimen if what we want to see is the external form of an object, presenting this to the eye in a way that is easy to comprehend because it appears solid: The transparency problem is avoided. Often directional lighting effects enhance the realism by adding the extra cue of light and shadow. Figure 11.11 shows an example where

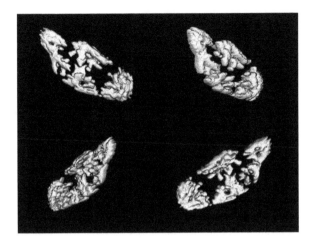

FIGURE 11.11   Chloroplasts in the dinoflagellate alga *Dinophysis*. Surface extraction, rendered with a lighting model. This gives a very realistic view and solves the transparency problem, at the expense of discarding all internal information.

a clear shape—chloroplasts in an alga—makes an ideal candidate for this style of presentation.

TECHNIQUES THAT CREATE VIEWS DIRECTLY FROM INTENSITY DATA

Simple Projections

In cell biology, we are often concerned with more complex relationships than can be represented with surface extraction. When this is so, our reconstructions must make use of all the information in the original data set. Doing so hugely increases the requirements for memory and computer power, but these have increased enormously since 1987, when confocal microscopy first became widely available. The simplest approach is to project the 3D array from a given angle onto a plane as when creating a stereo pair (Figure 11.6). If we take the average of all the intensity values (voxels) along each line (Figure 11.12A) the result is an *average brightness projection* (Figure 11.13). This projection also provides an improvement in the signal-to-noise ratio, because we are taking the average of a large number of views. However, because so much of a fluorescent data set is black, we end up with a rather dim image, and we always need to rescale the contrast in the final projection. In an 8-bit image with a lot of black space this can cause posterization (Chapter 6). Also in complex images it may be difficult to distinguish objects because relatively dim components make a large contribution.

Alternatively, if we find and record the maximum intensity along each line, rather than taking the average (Figure 11.12B), we form a maximum brightness projection (Figure 11.14). Because any structure is at its maximum brightness when in best focus, this method has logic behind it, but it does not improve the signal-to-noise ratio. In complex samples, this method gives prominence to the brightest features,

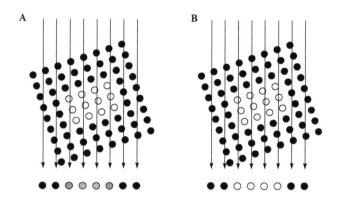

FIGURE 11.12 (A) Average brightness projection. As we follow the line down through the dataset we take the mean of the intensity values we encounter. (B) Maximum brightness projection. Following the line down, we store only the highest intensity we have encountered. As soon as a higher one is found we discard the previous value.

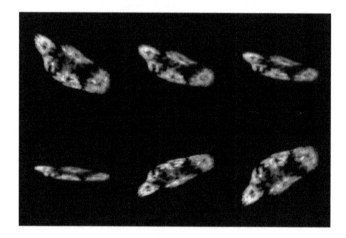

FIGURE 11.13 Six different views rotating around the X axis from an average brightness projection of the *Dinophysis* cell seen in Figure 11.11. Note that the transparency problem makes the same chloroplasts appear to be on top both in the real top view (upper left) and in the bottom view (lower right).

and dim ones may disappear. Which projection is more effective depends on the sample, but surprisingly often the results can be quite similar.

These projections simplify the computation, because we do not need to calculate the Z coordinates as we rotate the data set; it is enough to know where each point is in X and Y. Typically, we would keep either X or Y fixed and rotate about the other axis so that we are tracking translations in one direction only. This approach has the advantage that all internal detail contributes to the final image, but it suffers from the major problem that there is no distinction between front and back—both projections are identical. This problem can be alleviated if we create a

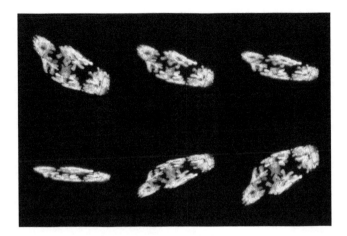

FIGURE 11.14   The same data set as Figure 11.13, from the same view angles, in a maximum brightness projection. The image looks noisier (because there is no averaging involved in the projection) but brighter. The transparency problem is the same, and the relatively poor Z resolution is also apparent in both projections.

stereo pair rather than a single image at each viewpoint. Nevertheless, if the image is complex, the problem of visualizing totally transparent objects may make the image hard to comprehend.

In Figure 11.13 and Figure 11.14 the instrumental problem of poor resolution in the Z direction is very apparent. In fact, the Z resolution appears much worse than it really is, which is a natural consequence of the confocal imaging process. As we saw in Chapter 5, the confocal pinhole never totally rejects light from an out-of-focus object; the response curve in Z gradually diminishes, with no minimum, so that objects just fade out. The eye likes to find edges (Chapter 6), and because there are no edges we perceive a much worse resolution than the full width at half maximum (FWHM) would suggest. A big improvement can be made by relatively simple image restoration techniques (Cox & Sheppard, 1999), as discussed in Chapter 10. Figure 11.15 shows the results of such processing on the same data set as in Figure 11.14, in the same maximum brightness projection. The original data set is shown in 11.15A. In 11.15B, it has been processed by first using a median filter (in three dimensions) to reduce noise, and then applying a one-dimensional restoration filter to sharpen the axial (Z) response and give it a clear minimum.

Weighted Projection (Alpha Blending)

In this technique, a projection is again created by summing all the voxels along each line of sight but with the added refinement that the voxels in front are given more importance than those behind (Figure 11.16). In this diagram, the group of white voxels on the right is near the top, so these voxels contribute strongly to the final intensity, whereas the similar group on the left is low down, so its voxels contribute less. In other words, opacity cues are reintroduced, but partially, so that internal detail is not lost. In this way, at each viewpoint the front is distinct from the back, yet the interior

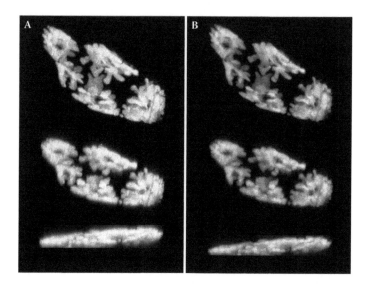

FIGURE 11.15   The same dataset of *Dinophysis* as Figure 11.13 and Figure 11.14 in maximum brightness projection, tilted at 0°, 45°, and 90°. (A) Original dataset; (B) after application of a 3D median filter and 1D deconvolution, showing much better resolution in the vertical axis.

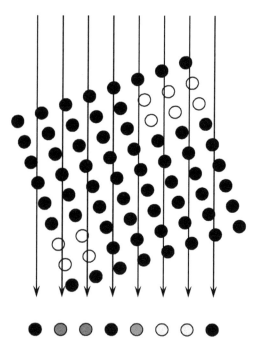

FIGURE 11.16   Alpha blending, a projection method in which bright structures located near the front (top in this diagram) are given more weight than ones at the back (bottom).

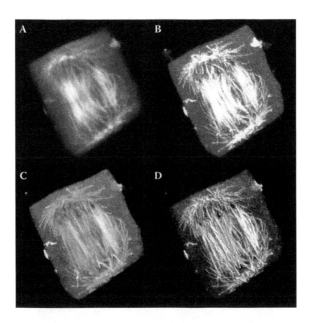

FIGURE 11.17 Different rendering methods applied to a confocal data set of a dividing wheat root cell. The specimen is stained with FITC anti-tubulin and shows the spindle microtubules at late anaphase. The data set is $256 \times 256 \times 88$, extracted from a larger field. In each case, the projection is a top view of the data, looking straight down the stack of sections with no rotation. All projections were scaled to a uniform contrast range. (A) An average brightness projection. (B) A maximum brightness projection. The average brightness projection gives a good impression of the overall distribution of microtubules, but with very little impression of depth; a view from the opposite direction would be identical. The maximum brightness projection gives much more prominence to individual microtubules and small groups. However, noise is much more apparent because there has been no averaging. (C) A weighted (or alpha-blended) reconstruction. The tonal range gives a much clearer indication of what is near the top and what is farther down the stack. (D) A weighted projection with a lighting model applied. The individual microtubules (or bundles of microtubules) are much easier to distinguish, thanks to the shadows cast.

of the sample remains visible (Figure 11.17C). Figure 11.17A and Figure 11.17B show corresponding average and maximum projections for comparison.

Weighted projection not only has the overhead of computing the transparency, but it also requires the Z position of each voxel be calculated at each rotation angle, which adds to the computational demands. Until recently, this type of reconstruction was the preserve of expensive software running on equally expensive computers. Now, a normal personal computer can handle the task, though it will need plenty of RAM (random-access memory); as a rule of thumb, enough to hold the entire data set twice over. When the hard disk must be used, performance slows dramatically. Commercial software remains expensive, but freeware and open-source utilities now exist as an alternative.

In a weighted reconstruction, the level of transparency allocated to a voxel is based on its brightness, but the exact relationship is under the user's control, and

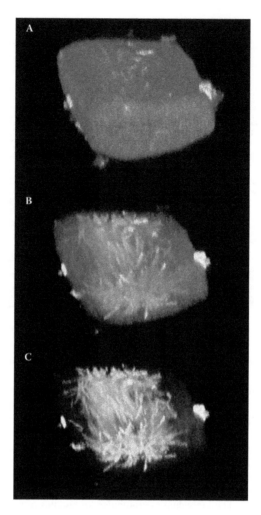

FIGURE 11.18   Transparency. The same data set as Figure 11.17 in an oblique view. (A) If we keep even midlevel voxels fairly opaque, we see mostly the cytoplasm of the cell and little of the contents. (B) Bringing opacity closer to linear, we begin to see the spindle. (C) Setting the opacity so that mid- to low-intensity voxels have little effect shows the spindle almost exclusively.

in practice needs considerable experimentation to give the best reconstruction (Figure 11.18; Chen et al., 1995). The result is inevitably a compromise between seeing internal detail and giving a clear front–back differentiation, but it can be a very effective compromise, and we can generate as many reconstructions as we want. In most cases, we probably want to create these reconstructions as rotation sequences that we can play as a movie, bringing the visual cue of motion parallax into play.

Most software offering weighted projection can also add a lighting model to the reconstruction. In this case, each voxel is treated not as an intensity value but as a reflective point onto which light shines from a source the user can position (Chen et al., 1995). This generates shadows, which can make surface detail much clearer

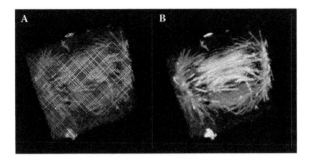

FIGURE 11.19 Aliasing. (A) At certain orientations, aliasing can be extremely distracting, making the image almost invisible. (B) Antialiasing software solves the problem at the expense of a substantial increase in rendering time.

and help with front–back relationships (Figure 11.17D). Lighting also strongly emphasizes noise, which appears as granular detail on the surface; do not mistake this noise for true surface structure. If noise is obtrusive, consider preprocessing the data with a smoothing filter (Chapter 10; see also Figure 11.15). Applying a lighting model adds hugely to the processing requirements and will slow things down, even on a fast computer.

When a data set is rotated, some positions cause a sampling line to hit a much higher number of voxels than its neighbor. When this happens, the reconstruction shows alternating dark and bright stripes in a parallel or crisscross pattern (Figure 11.19A). This problem, known as *aliasing,* can be avoided by treating each voxel as a cube in space rather than as a point (Figure 11.19B). Although this technique better approximates the truth, it involves another large increase in the amount of processing required, slowing the generation of the reconstruction. In general, experiment first to obtain all the correct parameters for your reconstruction; turn on luxury features, such as antialiasing and lighting, only to generate the final view or series of views.

If we create a movie from a series of views generated by a weighted projection algorithm, we have recovered the two key visual cues of motion parallax and opacity (objects in front obscuring those behind). The light and shade cue can likewise be regained by creating a reconstruction with a lighting model. If we create the movie as a series of stereoscopic views, we also bring in binocular parallax. In addition, some software offers perspective, introducing a scaling difference between objects at the front and the back (and, of course, making further demands on computing power). All the visual clues that we are ever likely to use in reconstructions are now available to us. The ones we cannot get (convergence, focus) remain out of our reach as long as computers produce images on two-dimensional screens. Will we have holographic displays by the time the next edition of this book appears?

As 3D reconstructions become more powerful, the problem of poor Z resolution in confocal data sets becomes more serious: Images become fuzzier as they rotate toward a side-on view. It seems likely that the sort of preprocessing shown in Figure 11.15 will be used more widely in the future. On the other hand, the rapid progress in super-resolution microscopy (Chapter 17) may make any sort of deconvolution unnecessary.

REFERENCES

Chen, H., Swedlow, J.R., Grote, M., Sedat, J.W., and Agard, D.A. 1995. The collection, processing and display of digital three-dimensional images of biological specimens. In *Handbook of Biological Confocal Microscopy*, 2nd ed., J.B. Pawley, ed. New York: Plenum Press, pp. 197–201.

Cline, H.E., Lorensen, W.E., Ludke, S., Crawford, C.R., and Teeter, B.C. 1988. Two algorithms for the three-dimensional reconstruction of tomograms. *Medical Physics*, 15, 320–327.

Cox, G.C., and Sheppard, C.J.R. 1999. Appropriate image processing for confocal microscopy. In *Focus on Multidimensional Microscopy*, vol. 2, P.C. Cheng, P.P. Hwang, J.L. Wu, G. Wang, and H. Kim, eds. Singapore: World Scientific Publishing, pp. 42–54.

12 Green Fluorescent Protein

In 1962 Osamu Shimomura, Frank Johnson, and Yo Saiga published a paper describing the purification of *aequorin*, a calcium-sensitive, chemiluminescent protein responsible for the bioluminescent properties of the jellyfish *Aequoria victoria* (Shimomura et al., 1962). When excited, this jellyfish emits green light at the fringes of its tentacles. Pure aequorin gives off a blue bioluminescence (emission peak 470 nm), but Shimomura and Johnson also found a companion protein "exhibiting a very bright, greenish fluorescence," now called by the descriptive, if not very original, name of *green fluorescent protein* (GFP). This protein had an excitation peak close to 470 nm, suggesting that that the blue emission given off by aequorin is transferred to GFP via resonant energy transfer (Chapter 15) and GFP, in turn, emits the green glow characteristic of *Aequoria*.

A fluorescent protein (FP) was an unusual concept at that time, which naturally invited further investigation. It was found that GFP was encoded by a single gene, which was in due course sequenced. This was the beginning of a revolution in cell biology. GFP proved to be a very stable protein; its fluorescence is unaffected by harsh conditions such as 8 M urea, 1% SDS, and 2 days' treatments with various proteases; it also survives aldehyde fixation. GFP is stable over a large range of pH (5.5 to 12) and at temperatures up to 65°C. Because it is so stable, is encoded by a single gene, needs no posttranscriptional modification, and requires no cofactors, GFP has revolutionized cell biology by providing an expressible marker that is directly visible in the fluorescence microscope (Chalfie et al., 1994; Inouye & Tsuji, 1994).

STRUCTURE AND PROPERTIES OF GFP

GFP is a small (27 kDa) protein consisting of 238 amino acid residues in a single polypeptide chain that forms an 11-stranded drumlike structure, termed a β-can, with an alpha-helix running up its center (Figure 12.1; Ormö et al., 1996, Yang et al., 1996). The can, or barrel, has the approximate dimensions of 3 nm in diameter and 4 nm in height. The actual chromophore is formed from just three amino acids: Ser65-Tyr66-Gly67. On its own, this structure is not fluorescent; the fluorophore is created by the autocatalytic oxidation of three peptides in the central α-helix to give a phenolic ring structure (Figure 12.2). GFP's properties are further modified by hydrogen bonding between the fluorophore and the surrounding β-can. This maturation of GFP is posttranslational and autocatalytic, requiring no other enzymes or cofactors.

The β-can structure provides a protected environment for the chromophore, which accounts for its extraordinary stability. The chromophore is shielded from oxygen (and therefore the fluorescence is very resistant to bleaching) and denaturing agents. The can is also essential for controlling the configuration of the three amino acids that form the fluorophore: Attempts to create a fluorescent peptide without the can have met with little success.

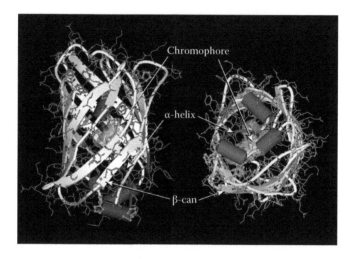

FIGURE 12.1 The structure of GFP. (Left) Side view; (right) top view. (Courtesy of Alexander Savitsky.)

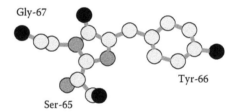

FIGURE 12.2 The fluorophore of GFP.

Wild-type GFP has a major excitation peak at 395 nm and a minor one at 475 nm (Figure 12.3). It has an emission peak at 509 nm. This unusual spectrum is explained by the existence of two isoforms: One absorbs at 395 nm but does not fluoresce; the other absorbs at 470 nm and fluoresces at 508 nm. The two forms interconvert in what is called a *Förster cycle* (Figure 12.4). The existence of this cycle also explains why wild-type GFP, even though it has a high quantum efficiency (most photons absorbed do give rise to fluorescent photons), has a rather low extinction coefficient (it does not trap many photons). Only the protonated fraction of the population (Figure 12.4, upper left) is efficient at absorbing light (Tsien, 1998).

GFP VARIANTS

Revolutionary it may have been, but wild-type GFP (wtGFP) has several deficiencies as an expressible probe, which, by definition, will be used mostly on living cells. In this regime, light dose is critical, and the low extinction coefficient is undesirable, as is excitation in the ultraviolet (UV) region, where cytotoxicity is a problem. The delay between protein expression and fluorescence, while the oxidation that forms

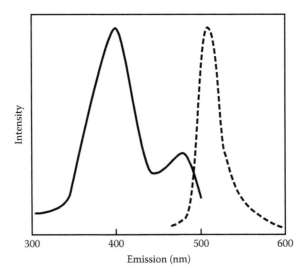

FIGURE 12.3 Excitation (solid) and emission (dashed) spectra for wild-type GFP.

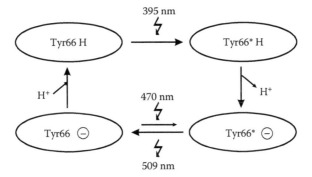

FIGURE 12.4 The Förster cycle of wild-type GFP. With tyrosine 66 in its hydroxyl state (upper left) the molecule can absorb a photon at 395 nm, taking it to an excited form (upper right), which loses a proton taking it to its phenolate form (lower right). This is the only form that fluoresces, emitting a photon at 509 nm and returning to a nonexcited state (lower left). This picks up a proton, returning it to the hydroxyl state and completing the cycle. Less often, it will absorb a photon itself at 470 nm, returning to the excited phenolate state.

the fluorophore takes place, poses problems for some experiments, particularly dynamic studies of gene expression. GFP fluoresces less well at temperature above 25°C, temperatures never encountered in the chilly waters in which *A. victoria* is found but which are essential for mammalian cell studies.

Many laboratories have sought amino acid substitutions through mutagenesis that lead to more efficient and useful GFP variants. For example, reconfiguring the molecule to be permanently in an efficiently excited "470 nm" state provides a higher extinction coefficient and a more useful excitation wavelength. The term *eGFP* is

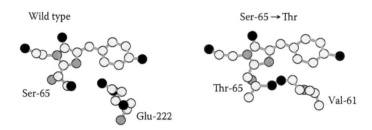

FIGURE 12.5 A single amino acid substitution from wtGFP shifts the major absorbance peak to 490 nm and increases the fluorescence intensity 30-fold. (Data from an unpublished honors thesis by Matthew Georgy, 1999.)

used rather generically to refer to such versions. A huge number of GFP variants have been created, some seeking to improve efficiency, others providing alternative excitation and emission wavelengths for multiprobe studies.

A further problem with wtGFP is that expression can be poor in some systems. Here, the solution has come from changing to alternative DNA triplets encoding the same amino acids, giving DNA sequences that are, for example, more readily expressed in mammalian cells. Plant systems in particular often express wild-type GFP inaccurately, so plant-specific versions of the gene have also been generated. There is also a cryptic intron in wild-type GFP that is removed in some variants, again making for more accurate expression in some systems.

Figure 12.5 shows one example of the kinds of changes that can be made: The single amino acid substitution Ser65-Thr changes the conformation of the fluorophore so that excitation is at the more cell-friendly wavelength of 490 nm and greatly increases the fluorescence yield. Many spectrally shifted variants have been produced, allowing multiple tagging and more efficient uses of available light sources. For example, another single substitution, Tyr 66 → His, shifts the emission to the blue and the excitation to the near-UV. Substitutions at amino acids 163 and 175 (which are in the β-can, not the fluorophore) improve fluorescence at mammalian temperatures. Typically these modified forms of wild-type GFP have few (<10) amino acid substitutions.

The next revolution came in the closing years of the 20th century when it was realized that a wide range of other jellyfish relatives (cnidarians) also possess similar fluorescent pigments. The first of these was DsRed, discovered by Mikhail Matz in the sea anemone *Discosoma* (Matz et al., 1999). DsRed is only about 30% homologous with GFP but has the same β-can structure. It is a tetramer, which can make it unwieldy in some applications. Another disadvantage is that it requires a sometimes lengthy postexpression maturation to develop the red fluorescence: When first expressed, it fluoresces in the green. However, DsRed extended the color range very usefully. The discovery of DsRed proved to be the tip of the iceberg. Corals had long been known to have fluorescent pigments, and now these were recognized as also part of the GFP family. There is now known to be an enormous variety of these pigments. Figure 12.6 shows four that have been isolated from one single species of coral (which also houses a nonfluorescent chromoprotein belonging to the same family).

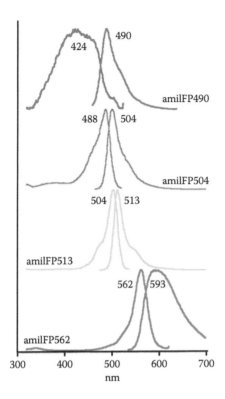

FIGURE 12.6 Spectra of four FPs, spanning the blue to the red, isolated from just one species of coral, *Acropora millepora*. (From Cox, G.C., Matz, M., and Salih, A., 2007, *Microscopy Research & Technique*, 70, 243–251. With permission.)

Coral pigments have many advantages over *Aequoria* derivatives. For example, the blue fluorescent versions are much more stable than blue derivatives of GFP, and there is more choice at the red end of the spectrum. All have the β-can structure and are typically 30% to 40% homologous with wild-type GFP. Most of the coral proteins (at least those investigated so far) seem to be tetramers, and mutagenesis to develop derivatives from these proteins has focused on creating monomeric forms, which retain the spectral diversity, and faster maturing forms. Figure 12.7 shows the spectra of just some of the many varieties of fluorescent protein currently available.

"We happened to leave one of the protein aliquots on the laboratory bench overnight. The next day, we found that the protein sample on the bench had turned red, whereas the others that were kept in a paper box remained green. Although the sky had been partly cloudy, the red sample had been exposed to sunlight through the south-facing windows."

Thus began the next revolution, in a paper from Ryoko Ando and others (Ando et al., 1999) from Atsushi Miyawaki's lab in Japan. They had extracted a fluorescent protein from the coral *Trachyphllia geoffroyi* (purchased from a local aquarium supplier) and found it to be excited in the blue at around 490 nm, with green

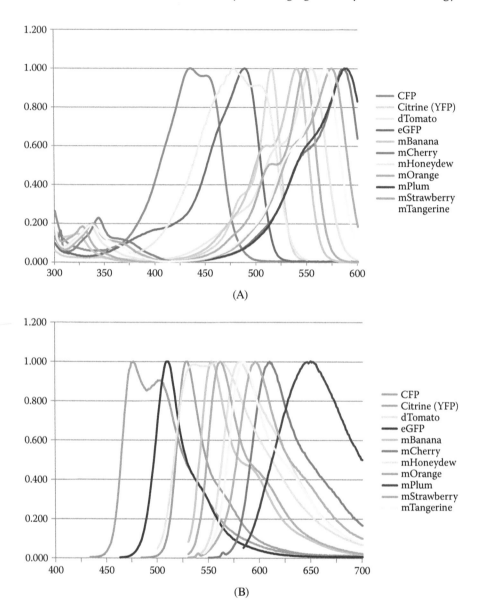

(A)

(B)

FIGURE 12.7 (A) Excitation and (B) emission spectra of 11 from the many FP variants now available. m = monomeric protein, d = dimer. (Data courtesy of Roger Tsien and Paul Steinbach, from http://www.tsienlab.ucsd.edu/.)

fluorescence at around 520 nm. Like most members of the GFP family, its fluorescence seemed stable and long lived. But excitation in the near-UV transformed it into an equally stable form fluorescent in the red, with its excitation maximum in the green. They nicknamed this protein Kaede, from the Japanese word for "maple," because the color change resembled maple leaves in autumn.

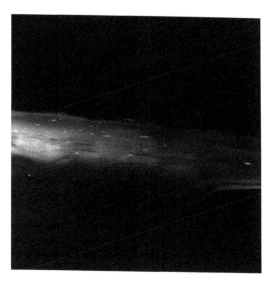

FIGURE 12.8 Purified EosFP in 1% agarose, dried, photoconverted to red in the center by 30s irradiation at 340–380 nm (epifluorescence illuminator).

Many other photoactive FPs have since been isolated. Figure 12.8 shows Eos (Greek for "dawn"), which has a green to red conversion like Kaede and comes from the coral *Lobophyllia hemprichii* (Wiedenmann et al., 2004). A Russian group found that a sea anemone GFP homologue, which had little or no fluorescence in its native state, could be activated by intense light and then became stably fluorescent (Lukyanov et al., 2000). These are now known as kindling proteins. In fact, kindling had been observed in wild-type GFP, but the effect was not sufficiently large or stable to offer a practical use. Miyawaki's lab then found Dronpa (from the Japanese for "appear and disappear"), which is normally excited in the blue. Green light would destroy the fluorescence, but near-UV would restore it, and the process could be repeated for hundreds of cycles without any sign of damage (Ando et al., 2004). Reversible color-changing forms are also now being discovered. These convertible proteins offer interesting possibilities in following cell dynamics. Selective photobleaching in FRAP (Chapter 14) always raises the possibility that the light levels used are cytotoxic, but a photoactive protein can be used for similar experiments with only the light levels needed in any fluorescent microscopy (admittedly, even these are not always kind to cells).

APPLICATIONS OF GFP

GFP–protein fusions can be done at either the C- or the N-terminus of GFP without inhibiting the fluorescence. GFP has been successfully tagged to a multitude of proteins, including cytoskeletal-associated proteins and organelle-specific proteins in plants, animals, and yeasts. It may be transfected into the host and expressed permanently as a stable part of the genome or a suitable DNA or RNA construct can be put into a cell where (with luck) it will be transcribed; this is called *transient expression*.

A multitude of techniques exist for transfection, some chemical and others mechanical. This section looks at some of the methods for both permanent and transient transfection from the perspective of what can be done with each.

HEAT SHOCK

Plasmids are short strands of DNA that are carried and expressed by bacteria. Even though they do not become incorporated into the (single) bacterial chromosome, they replicate and are passed on to daughter cells. Plasmids play an important role in many bacteria, and incorporation and transfer of plasmids is part of a bacterium's natural life cycle. It is therefore not difficult to get a plasmid into a bacterial culture. Transferring the culture from ice to 42°C (for less than a minute) and back is sufficient to make bacteria take up external plasmids. Typically, we include in the plasmid, besides our FP construct, genes to enable the bacteria to survive stringent conditions (antibiotic resistance is a simple example), so that we can then isolate those that have taken up the plasmid.

CATIONIC LIPID REAGENTS

The commercial cationic lipid reagents Lipofectin and Lipofectamine (and many other derivatives) form submicron unilamellar liposomes in water when dispersed under appropriate conditions. The exterior of the liposomes is positively charged and is therefore attracted electrostatically to the phosphate backbone of DNA. The DNA does not enter the liposomes, but remains external; in fact, it seems that several liposomes may attach to one DNA strand.

The cell membrane also typically carries a negative charge on its surface, so the liposomes plus their DNA cargo become attached to the cell surface. From there they can be taken up, probably more or less accidentally, by endocytosis.

This approach primarily applies to transfecting cultured cell lines, but that is often a prerequisite to transfection of whole organisms.

DEAE–DEXTRAN AND POLYBRENE

DEAE–dextran and polybrene work on a similar basis to cationic reagents, but in this case the positive charge is carried by the large polymer molecules. The positively charged DEAE–dextran or polybrene molecules bind electrostatically with negatively charged DNA molecules and also to the cell surface, thereby enabling the DNA to bind to the cell surface. Uptake of the complexed DNA into the cell is accomplished by osmotic shock using DMSO or glycerol. Both reagents have been used successfully for DNA transfections of cultured cell lines, although DEAE–dextran is limited to use in transient transfections.

CALCIUM PHOSPHATE COPRECIPITATION

Mixing of calcium chloride, DNA, and phosphate buffer precipitates extremely small, insoluble particles of calcium phosphate containing condensed DNA. The calcium phosphate–DNA complexes adhere to cell membranes and enter the cyto-plasm of the target cell by phagocytosis. The size and quality of the precipitate are crucial factors in the success or failure of $Ca_3(PO_4)_2$ transfections.

ELECTROPORATION

Electroporation involves exposing cell cultures to a brief electrical pulse of high field strength. The electric field creates a potential difference across the membranes of individual cells, which is believed to induce temporary pores in the cell membrane. This enables us to deliver molecules into cells and is used in staining (Chapter 13) as well as in transfection. Optimization of the electrical pulse and field strength param-eters is essential for successful transfections, because excess irreversibly damages the membrane and lyses the cells. Even so, high levels of toxicity (50% or greater) generally accompany successful electroporation procedures.

MICROINJECTION

Transfection technology can also use mechanically based methods such as *micro-injection*. Microinjection introduces DNA or RNA directly into the cell cytoplasm or nucleus using a very fine micropipette under a microscope. This technique is technically demanding and time consuming but much less hit-or-miss than other approaches. If the aim is to create an entire transgenic animal, microinjection is almost inevitably part of the process. In this case, the usual approach is to microin-ject a plasmid, or even whole chromosomes from a successfully transfected cell line, into an ovum.

GENE GUN

A crude but amazingly effective technique for transient transfection is to use a *gene gun*, a device that fires colloidal gold particles coated with DNA or RNA into tissue at high velocity. The diagram in Figure 12.9 is loosely based on the Helios gene gun from Bio-Rad, but there are others and many labs have built their own. The basic

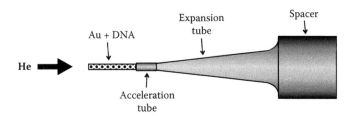

FIGURE 12.9   A gene gun. A high-pressure blast of helium accelerates colloidal gold par-ticles through the acceleration tube. In the expansion tube, the helium spreads out and loses its velocity, but the gold continues at high speed. The spacer is selected to determine the area over which the gold will spread. (Diagram based on the Bio-Rad Helios design.)

principle of a gene gun is to use a high-velocity burst of helium to accelerate gold particles coated with plasmids or RNA to velocities sufficient to penetrate cells, particularly cells and tissues in vivo. In the Helios system, the gold particles are dried on the walls of a 1.5-mm tube (the cartridge). Helium travels down the bore of the particle delivery device; when it enters the cartridge, the gold particles on the inside of the tubing are pulled from the surface, become entrained in the helium stream, and begin to pick up speed. Immediately past the acceleration channel, the barrel opens as a cone, expanding the high-pressure jet into a less destructive low-velocity pulse, while the dense gold particles maintain their high velocity. The expansion also helps spread the gold microcarriers out from their original 1.5 mm diameter to bombard an area approximately 12 mm in diameter at the target site. The size of the spacer affects this: the longer the spacer, the larger the area. Variants have the gold suspended in liquid, with the expectation that it will be vaporized—or at least atomized—by the time it reaches the target. The initial helium pressure is the other important variable: high for skin, tough tissues, and plant cells with cuticles; much lower for soft tissue.

Figure 12.10 shows an *Impatiens* (balsam) epidermal cell transfected using the Helios gene gun; the construct is GFP conjugated to h-del, a signal peptide that acts as an instruction for sequestration in the ER, so that the GFP accumulates in the ER.

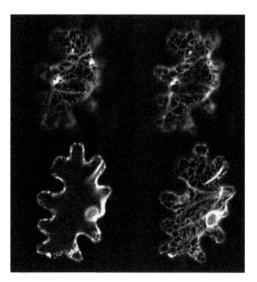

FIGURE 12.10 Four planes through a living *Impatiens* epidermal cell after transformation by bombarding the leaves, on their lower side, with 1.6 µm gold particles coated with plasmid DNA, containing the *mgfp5*-ER construct from Jim Haseloff's lab (Siemering et al., 1996). In this construct the fusion of the N-terminal signal peptide sequence from *Arabidopsis thaliana* basic chitinase and the C-terminal HDEL sequence (ER retention signals) allows compartmentalization of GFP into the lumen of the endoplasmic reticulum.

FIGURE 12.11   Epidermal cells of *Arabidopsis thaliana* permanently expressing GFP–β-tubulin (green). The red bodies are chloroplasts made visible by autofluorescence of the chlorophyll.

PLANTS: AGROBACTERIUM

Permanent transfection in plants is invariably carried out using *Agrobacterium tumefaciens*, a common soil bacterium that causes crown gall disease by transferring a particular DNA segment (T-DNA) of the tumor-inducing (Ti) plasmid into the nucleus of infected cells of the plant host. The transferred DNA is stably integrated into the plant genome, where its expression leads to the synthesis of plant hormones and thus to the tumorous growth of the cells. Genes for the synthesis of opines—compounds on which the bacterium feeds—are also transferred.

The T-DNA genes are transcribed only in plant cells, not in the bacterium, and do not play any role in the transfer process. Expression of the T-DNA genes in the plant causes the tumor to form. Because the bacterium has no control over the T-DNA section of the plasmid, it follows that any foreign DNA placed between the T-DNA borders will be transferred to the plant cell, wherever it comes from.

As long as the regions of the plasmid that control infection are unaltered, we have a lot of freedom as to what we can do within the T-DNA region. We can include the genetic payload of our choice and inactivate tumor formation without affecting the infection process. Figure 12.11 shows one example: leaf cells in a plant of *Arabidopsis thaliana* that is stably expressing GFP-tagged β-tubulin. The transfection is stable and permanent, and the transgenic plants can readily be grown from seed.

REFERENCES

Ando, R., Hama, H., Yamamoto-Hino, M., Mizuno, H., Miyawaki, A. 1999. An optical marker based on the UV-induced green-to-red photoconversion of a fluorescent protein. *Proceedings of the National Academy of Science USA*, 99, 12651–12656.

Ando, R., Mizuno, H., and Miyawaki, A. 2004. Regulated fast nucleocytoplasmic shuttling observed by reversible protein highlighting. *Science*, 306, 1370–1373.

Chalfie, M., Tu, Y., Euskirchen, G., Ward, W., and Prasher, D., 1994. Green fluorescent protein as a marker for gene expression. *Science*, 263, 802–805.

Cox, G.C., Matz, M., and Salih, A. 2007. Fluorescence lifetime imaging of coral fluorescent proteins. *Microscopy Research & Technique*, 70, 243–251.

Inouye, S., and Tsuji, F. 1994. *Aequorea* green fluorescent protein. Expression of the gene and fluorescence characteristics of the recombinant protein. *FEBS Letters*, 341, 277–280.

Lukyanov, K.A., Fradkov, A.F., Gurskaya, N.G., Matz, M.V., Labas, Y.A., Savitsky, A.P., Markelov, M.L., et al. 2000. Natural animal coloration can be determined by a non-fluorescent green fluorescent protein homolog. *Journal of Biological Chemistry*, 275, 25879–25882.

Matz, M.V., Fradkov, A.F., Labas, Y.A., Savitsky, A.P., Zaraisky, A.G., Markelov, M.L., and Lukyanov, S.A. 1999. Fluorescent proteins from nonbioluminescent Anthozoa species. *Nature Biotechnology*, 17, 969–973.

Ormö, M., Cubitt, A.B., Kallio, K., Gross, L.A., Tsien, R.Y., and Remington, S.J. 1996. Crystal structure of the *Aequorea victoria* green fluorescent protein. *Science*, 273, 1392–1395.

Shimomura, O., Johnson, F.H., and Saiga, Y. 1962. Extraction, purification and properties of aequorin, a bioluminescent protein from the luminous hydromedusan, *Aequorea*. *Journal of Cellular and Comparative Physiology*, 59, 223–239.

Siemering, K.R., Golbik, R., Sever, R., and Haseloff, J. 1996. Mutations that suppress the thermosensitivity of green fluorescent protein. *Current Biology*, 6, 1653–1663.

Tsien, R.Y., 1998. The green fluorescent protein. *Annual Review of Biochemistry*, 67, 509–544.

Wiedenmann, J., Ivanchenko, S., Oswald, F., Schmitt, F., Röcker, C., Salih, A., Spindler, K.-D., and Nienhaus, G.U. 2004. EosFP, a fluorescent marker protein with UV-inducible green-to-red fluorescence conversion. *Proceedings of the National Academy of Science USA*, 101, 15905–15910.

Yang, F., Moss, L.G., and Phillips, G.N. Jr. 1996. The molecular structure of green fluorescent protein. *Nature Biotechnology*, 14, 1246–1251.

13 Fluorescent Staining

Teresa Dibbayawan, Eleanor Kable, and Guy Cox

IMMUNOLABELING

Immunolabeling uses the body's own immune response to generate highly specific probes. To understand the strengths and the pitfalls of this approach we need a basic knowledge of the workings of the immune system.

The traditional view of the immune system was developed some 70 years ago by Sir Frank Macfarlane Burnet (1957); in 1960, Burnet became the first Australian to be awarded a Nobel Prize. In Burnet's scheme, there are two broad classes of immune response: cell-mediated responses and antibody responses. *Cell-mediated responses* involve the production of specialized cells, the cytotoxic T lymphocytes, that bind to and lyse foreign or infected cells. *Antibody (humoral) responses* involve the production of *antibodies* by B-lymphocytes, which circulate through the body. Antibodies can bind and inactivate the response-evoking *antigens* (antibody generators) or mark them to be destroyed either by phagocytosis or in a complement-mediated process. It is the antibody response that we use in immunolabeling.

All lymphocytes are derived from the bone marrow, but T cells undergo a process of maturation in the thymus gland. There are two types of T cells: cytotoxic T cells (described earlier), which do not concern us further in this chapter, and *helper T cells,* which stimulate activation of the B lymphocytes. B lymphocytes are produced and mature in the bone marrow. (*B* originally stood for *bursa of Fabricius*, the specialized organ that produces B-lymphocytes in birds—most early immunology was done on chickens. It was perhaps fortunate that in mammals B lymphocytes also are produced in an organ whose name begins with *b*.)

Mature lymphocytes all have a similar appearance: small, nondescript cells with a deeply basophilic nucleus and scanty cytoplasm. They circulate in the blood and through body tissues. Each carries on its surface one specific antibody (which, in the case of a B cell, is the antibody it will eventually produce). There are about 10^7 possibilities for these antibodies, all formed during embryonic development, so the system can recognize a predefined repertoire of some millions of antigens. Antibodies that recognize self proteins are eliminated in the embryo. (Modern research has shown that in addition to this classical system there is also an adaptive system that can recognize novel antigens not catered for in the embryonic repertoire.)

Once a lymphocyte binds to its particular antigen, a chain of events begins, leading to massive production of the T and B cells of the appropriate clone, recognizing the stimulating antigen. The B cells become activated and differentiate into *plasma cells*, antibody-secreting cells with a large cytoplasm and masses of ER. These take up residence in the secondary lymphoid organs, particularly lymph nodes and spleen. Each B cell clone makes antibody molecules with a unique antigen-binding site, and these will soon circulate in large quantities in the blood.

Residual antibodies that circulate in the blood after the immune response is over provide the animal with an immunological "memory," whereby a more rapid and stronger response is mounted on second or subsequent exposures to a particular antigen.

TYPES OF ANTIBODY

Antibodies (immunoglobulins) are glycoproteins, formed from two identical heavy chains and two identical light chains, arranged in a Y shape (Figure 13.1). The four polypeptide chains are held together by disulfide bridges and noncovalent bonds. Figure 13.1 shows them in diagrammatic form; in reality, they are twisted round each other. The carboxyl ends of the heavy (i.e. long) chains (the tail of the Y) are the point of attachment when antibodies are presented at the surface of lymphocytes. Close to the amino terminal of each chain (i.e., at the tips of the arms of the Y) is a highly variable region that produces the antigen binding site, two per immunoglobulin molecule, each made up of one heavy and one light chain.

Proteolytic enzymes papain and pepsin can split an immunoglobulin into different characteristic fragments (Figure 13.1). *Papain* cleaves the Y-shaped molecule at the fork's three parts. The arms give us two separate and identical *Fab*s (fragments with antigen-binding sites), each with one antigen-binding site. The tail gives one

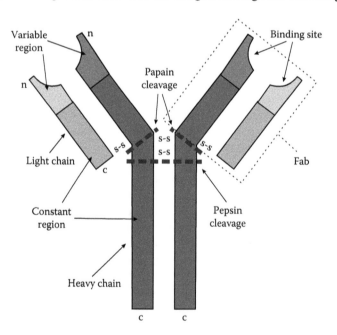

FIGURE 13.1 Diagram of an antibody (immunoglobulin) molecule. Two identical heavy (long) chains, and two identical light (short) chains, held together by disulfide bonds, form the basic immunoglobulin structure. *n* = amino terminal; *c* = carboxy terminal.

Fc (fragment that crystallizes) fragment. *Pepsin* cleaves the tail below the disulfide bridges, producing a single, V-shaped, antigen-specific fragment, but with two binding sites: $F(ab')_2$. The rest of the molecule's "tail" is broken into smaller pieces. Fab fragments represent the smallest component with antigenic activity and, therefore, the smallest unit we can use in fluorescent labeling.

Of the many different types of immunoglobulins, five are most important (and two are useful in immunolabeling). These five classes (IgG, IgM, IgA, IgE, IgD) are distinguished by specific heavy chains (γ, μ, α, ε, and δ, respectively). There are two types of light chains (κ and λ), either of which can be associated with any class of heavy chain and which do not appear to influence the properties of the antibody.

The five classes of immunoglobulins are as follows:

IgAs are dimers of the basic Y and are found in tears, saliva, and generally on mucous membranes where they provide portal resistance to infection.

IgDs are found in blood and lymph and are involved in the activation of B cells. They are found only in small quantities.

IgEs, the rarest, are involved in allergies and the inflammatory response.

IgGs are the most numerous, making up around 75% of the total immunoglobulins. IgGs are produced in response to prolonged or repeated exposure to an antigen and are highly specific. (When a pathologist's report says that you have been infected by something but are not infected any longer that statement is based on the presence of the appropriate IgG in your blood.) IgGs give us long-term resistance to a disease; they are also the antibodies of choice for immunolabeling, because they are monomeric and highly specific.

IgMs are pentamers and are produced on first exposure to an antigen. To a pathologist, they indicate a current infection. Typically IgMs are much less specific than IgGs, which fits their role as the frontline defense. Because of their size and low specificity, IgMs are less desirable for use in immunolabeling, but we may use them sometimes with so-called difficult antigens. (Difficult antigens are highly conserved molecules such as actin: There are rather few differences between host and foreign actin, so it does not elicit a strong immune response.)

RAISING ANTIBODIES

In cell biology we use antibodies to detect and localize features or molecules of interest in a cell. To raise an antibody, a group of animals is injected with an appropriate antigen that induces humoral responses in the animals. The initial injection produces mostly IgMs, but repeated injections over several weeks (if successful) produce a high level of IgGs in the bloodstream. These IgGs will not all be identical; the antigen stimulates a variety of lymphocytes, each of which produces a clone of antibody-secreting plasma cells, each responding to a specific site, or *epitope,* on the antigen, hence the term *polyclonal antibody.* Individual antibody molecules may have different binding affinities or see slightly different parts of the epitopes. Consequently, two polyclonal antibodies raised against the same antigen are never identical.

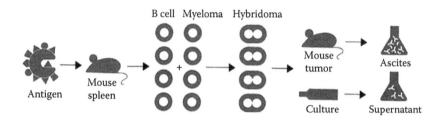

FIGURE 13.2   Monoclonal antibody production. (Courtesy of DAKO Cytomation Inc.)

A polyclonal antibody also, inevitably, contains immunoglobulins that were present prior to immunization. It is also likely to contain antibodies against any impurities in the prepared antigen. Purification prior to use is usually a must for polyclonal antibodies.

In 1975, Köhler and Milstein developed a technique that allowed for the growth of clonal cell populations secreting antibodies with defined specificity. Spleen cells from the immunized animal (almost always a mouse) are removed and mixed with cultured tumor cells. Membrane-disrupting reagents are used to promote cell fusion, and a small number of hybrid cells that remain immortal but produce antibody will result. Individual cells are plated and tested against the original antigen to isolate those that produce antibodies. The hybrid cells, *hybridomas*, are maintained in culture and continually secrete the monospecific antibody (Figure 13.2). For mass antibody production, these hybridomas may either be maintained in bulk cultures or reintroduced into a mouse, where they produce tumors, from which ascites fluid containing the antibodies may be drawn. Each of these *monoclonal* antibodies interacts with just one epitope, so they are highly specific. Anything from one to many different clones may result from one experiment, but each clone will have specific properties.

The choice between polyclonal and monoclonal depends on the application. Making monoclonal antibodies is laborious and expensive, but it does have the advantage that the antibody is then available effectively ad infinitum, and it will always be the same, guaranteeing reproducibility between experiments. Immunizing another animal to produce more polyclonal antibodies never gives exactly the same result. However, most applications involving the identification or intracellular localization of an antigen can usually be done effectively with good polyclonal antibodies. Because several epitopes are detected (rather than only one with a monoclonal antibody), more reliability and brighter fluorescence are generally obtained with the polyclonal serum. Monoclonals are highly specific. This can be an advantage where several related proteins are present, because polyclonals often show crossreactivity, but it is also possible that the epitope may be masked in the cell, so that the antigen is not detected. Monoclonal antibodies should be used when a specific epitope is of interest, as in epitope mapping. They can also be useful in identifying new components of a partially isolated cellular structure.

It is common practice to use a combination of polyclonal and monoclonal antibodies in dual labeling. The easiest choice is always to use commercially available antibodies, whether polyclonal or monoclonal, rather than raising an antibody yourself.

LABELING

To be visible in the confocal microscope, antibodies must have an identifiable molecule attached to them. Many markers have been used; horseradish peroxidase, which is easily demonstrated cytochemically with diaminobenzidine, has long been popular. Although this technique is valuable in histology, the precipitate is too diffuse to have much application in cell biology.

Fluorescent labeling is the norm when we need to look at subcellular levels. Typically, the dye (e.g., fluorescein, rhodamine), which is an acid, is converted to its isothiocyanate ester: $-N=C=S$. The isothiocyanate radical reacts with amine groups, so simply mixing the dye with the immunoglobulin conjugates the fluorochrome to the N-terminal (amine) group of the antibody chains. This is the antibody-binding end, which does raise the possibility of interference, but it also means that the fluorescent label ends up in very close proximity to the molecule of interest.

The only alternative label that is current is colloidal gold. Particles of colloidal gold bind electrostatically to proteins, just as they do to DNA (Chapter 12). Gold labeling is a routine approach in electron microscopy, because gold is highly visible in the electron image. In the optical microscope, the gold, although very reflective, is too small for easy detection, so the signal is amplified by silver precipitation. This gives a reflective signal that is easily seen in reflection confocal microscopy, or with epidarkfield illumination in the widefield microscope. When confocal microscopes typically offered only two fluorescence detection channels, gold labeling was often a handy way to image a third channel, but now that up to five detectors are available, the technique is less relevant. It is useful, however, if we want to look at a sample in both optical and electron microscopes, and it can also be handy if our tissue is very autofluorescent.

In principle, we could label our actual primary antibody with fluorescein isothiocyanate (FITC) or another fluorescent dye; this is called *direct labeling*. Direct labeling can give us very clean and specific labeling, but in practice we do not often use it—not least because we would have to start from scratch for each different antibody. More commonly, we incubate our tissue first with the specific antibody, and then with a labeled secondary antibody directed against the primary Ig; this is called *indirect labeling*. In this way, one labeled antibody (e.g., FITC anti-mouse) works with any antibody raised in the appropriate host (mouse, in this example). The brightness of labeling is greatly increased because many secondaries will attach to one primary IgG unit (Figure 13.3). Make sure, however, that your secondary

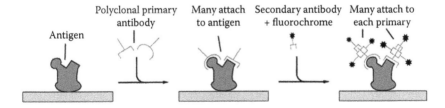

FIGURE 13.3 Indirect labeling.

antibody is directed against the species of your primary antibody. It is amazing how often this elementary point is forgotten.

Antibody *incubation times* and *dilutions* must be worked out empirically and methodically, using a series of dilutions (initially varied by factors of 10, i.e., 1:10, 1:100, 1:1,000; see Appendix C). It is also vital to carry out a range of controls for both primary and secondary antibodies. The best control is to use a *preimmune serum* from the same animal in which the antibody was raised or at least a non-immune serum from the same species if the preimmune is not available. This control will identify whether there is any nonspecific binding of your immunoglobulins. Other controls include leaving out one or the other or even both of the antibodies. If you get labeling even without the primary, there is nonspecific binding of your secondary antibody. If you get "labeling" without the secondary, that is, with no fluorochrome, it is clearly autofluorescence. It is also a good idea to add a *positive control*—an antibody you can trust to work—to verify that the labeling protocols are working, even if no signal was obtained from the particular antibody labeling. Commercial antibodies against tubulin are good for this purpose.

It is normally necessary to *block* any nonspecific binding of proteins that may take place in the specimen. If the sample has been fixed with an aldehyde, this blocking is absolutely essential because, by definition, aldehyde fixatives bind nonspecifically to proteins. The aim is to add saturating amounts of nonspecific proteins that will not be recognized by the secondary antibody, such as bovine serum albumen (BSA) or cold fish-skin gelatin (CFSG). Using the blocking buffer as the diluent for the antibodies may also be helpful.

Labeling with two or more primary antibodies requires primary antibodies raised in different hosts, so that each secondary "sees" the correct target. Figure 13.4 shows an example, with a commercial monoclonal antibody tagged with FITC anti-mouse and a rabbit polyclonal tagged with commercial Cy3 anti-rabbit. Dual labeling can

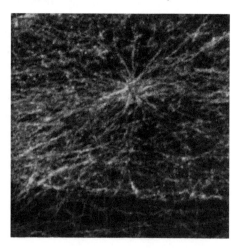

FIGURE 13.4 Microtubule organizing centers in cultured smooth muscle cells stained with commercial monoclonal anti-α-tubulin + FITC-anti-mouse and lab-raised polyclonal rabbit anti-γ-tubulin + Cy3-anti-rabbit.

be done simultaneously or sequentially, with fixation in between, depending on the suitability of the antibodies. The latter is safer as a starting point. Characterize and test each antibody in your system by single labeling before you attempt multiple labeling.

Modern dyes are often preferable to the traditional FITC (fluorescein isothiocyanate) and TRITC (tetramethyl rhodamine isothiocyanate). The AlexaFluor dyes from Molecular Probes (Life Technologies/Invitrogen), for example, offer greater photostability and more efficient conjugation, as well as a much wider selection of excitation and emission wavelengths. Details of all these dyes are given by Johnson and Spence (2011). It is common when labeling cytoplasmic structures to counterstain with DAPI or Hoechst (next section) to show the nucleus but do this with caution. These dyes have very broad spectra and can overwhelm weak fluorescence.

Several protocols that could be useful if you are just starting out are given in Appendix C. Of course, the best approach is finding someone who has already worked out what to do with your system. If this is not possible, read some literature and try different approaches until you find what works best with your material.

Because we normally work with fixed cells, it is a good idea to use antifade agents in the mountant media. These agents appear to work by scavenging oxygen, which easily reacts with fluorochromes when they are in an excited state (Chapter 3). The free radicals thus produced can damage even unexcited fluorochromes. Several commercial antifade reagents are available, such as Citifluor (Agar Scientific), ProLong (Molecular Probes), and FluoroGuard (Bio-Rad). The usual active ingredient is either p-phenylenediamine or n-propylgallate. It is best to view the labeled specimen soon after mounting but allow a little time for the antifade agent to penetrate through the specimen.

FLUORESCENT STAINS FOR CELL COMPONENTS AND COMPARTMENTS

A huge inventory of fluorescent probes is commercially available from a variety of sources (e.g., Molecular Probes/Invitrogen, Sigma-Aldrich). These probes target specific components or compartments of the cell. Some probes—*vital stains*—can be used on living cells; others are useful only for fixed material because they will not penetrate or are broken down by living cells. Some, such as the well-known nuclear stains DAPI and Hoechst, can be used either way but are more commonly used on fixed material. Vital dyes are particularly interesting because they enable us to visualize and quantify dynamic cellular events. We will not attempt to cover the range of probes now available, the *Molecular Probes Handbook* (Johnson and Spence, 2011) is the best source for that, but we offer a general overview of the types of probes and how they are introduced into the cell. Be sure to search the current literature, because new probes are produced daily.

Organelle probes are available for the mitochondria, endoplasmic reticulum, golgi, lysosomes, and nucleus. Most of these probes are cell-permeant stains, so you can label living cells by simply incubating the sample in the fluorochrome. Although earlier organelle probes suffered from lack of specificity, the new-generation probes are more specific and some have the advantage of being fixable in situ. For example, many of the MitoTracker® probes are fixable, whereas the older dye rhodamine 123 leaches out of the mitochondria on fixation. Both these probes are used in submicromolar solutions in appropriate buffers and incubated with the sample. The probes diffuse across the plasma membrane and accumulate in active mitochondria. The sample is then incubated in a control buffer to wash away excess dye, and cells are viewed under the microscope (Figure 13.5). The MitoTrackers contain a mildly thiol-reactive chloromethyl moiety. This appears to be responsible for retaining the dye associated with the mitochondria after fixation. The other advantage of the MitoTracker probes is that there is a range of them, from green to red, which allows you flexibility in planning your experiment.

The initial probe that was used as a marker of lysosomes was the metachromatic stain Acridine Orange (Figure 13.6). This dye lacks specificity; it also binds to DNA, where it fluoresces green, and to RNA, where it fluoresces red. Acidic polysaccharides also stain red and phenolics green. The newer Lysotracker® dyes consist of fluorophore of a biotin-linked moiety, linked to a weak base that is only partially protonated at neutral pH. These probes freely cross the cell membrane and concentrate in spherical organelles, the lysosomes. The probes, which must be used at low concentrations to achieve selectivity, come in a range of colors. Also, Lysosensor probes can detect changes in pH, showing a pH-dependent increase in fluorescence intensity upon acidification.

The ER and golgi are involved in the sorting of lipids and proteins in cells. Hence, most of the probes that are used to identify these organelles are either lipids or chemicals that affect protein trafficking. The short-chain carbocyanine dye $DiO_6(3)$ can be used to identify ER in both live and fixed cells. This dye passes through the plasma membrane and stains the intracellular membranes (Figure 13.7). This dye has two main disadvantages. It is not specific for ER, and one must use the characteristic

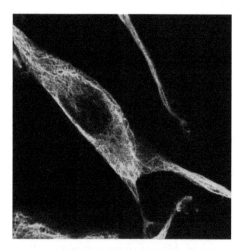

FIGURE 13.5   Cultured lung smooth muscle cells immunostained with FITC anti-α-tubulin and with mitochondria labeled with MitoTracker.

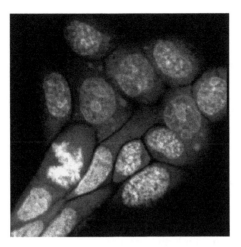

FIGURE 13.6 Confocal laser microscopic image of unfixed rat colon carcinoma cells in vitro incubated with 1 μg/L acridine orange for 10 minutes. The red fluorescent dots in the cytoplasm represent lysosomes. Besides the labeling of acidophilic subcellular compartments such as lysosomes, the nuclei (DNA) are stained green and the nucleoli (ribosomal RNA) red. Note at the left side a cell in division. Magnification, 630×, water immersion lens. (Courtesy of M. Timmers and F. Braet.)

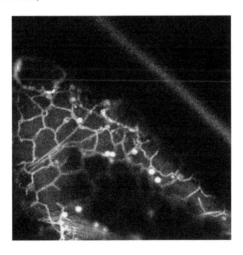

FIGURE 13.7   Endoplasmic reticulum in a living onion cell stained with $DiOC_6(3)$.

architecture of this organelle to distinguish it from others (e.g., mitochondria), and it is often toxic to the mitochondria before it stains the ER membranes. A further problem is that this dye does not come in a variety of colors, which is limiting in multilabel experiments. Fluorescent ceramide analogs are used to selectivity stain the golgi apparatus. These dyes have been used to study lipid transport in organelles.

The common probes used for staining the nucleus are DAPI and Hoechst 33258. DAPI (4′6-diamidino-2-phenylindole) is only semipermeant to live cells, so if you have used it in fixed cells, where the staining is incredibly bright, the fluorescence will seem very weak in living cells. These probes stain the chromosomes and nuclei. Hoechst can be used in live cell assays and will readily cross the cell membrane and bind to the chromosomes. These dyes bind to the minor groove of DNA at AT-rich sequences. Both of these probes have a very large emission range, so if you use them in conjunction with other probes you must be careful in choosing the other probes or be able to image your sample sequentially. The fact that DAPI and Hoechst need UV excitation also limits their usefulness in living cells, although this can be overcome by using two-photon excitation (Chapter 8). There are merits therefore in looking at longer-wavelength nuclear dyes. Invitrogen has a range of Syto nuclear dyes that are also cell-permeant and emit in the green-to-yellow fluorescent range, and the TOTO–TOPRO range, which emit in the far red.

Phalloidin and phallacidin are phallotoxins (toxins from the fungus *Amanita phalloides*) that can be fluorescently labeled and bind specifically to F-actin. These probes have the advantage of specificity, an ability to be used across a wide range of specimens, and an affinity for both large and small filaments. When using these dyes with live cells, though, you must be aware that the phallotoxin may alter the process you are studying as the actin's monomer–polymer equilibrium is shifted toward the polymer. These probes are also highly toxic, requiring a delicate balance between maintaining viability and getting usable staining. Even in fixed cells, phalloidin has been demonstrated to cause aggregation of actin filaments.

Lipophilic carbocyanine dyes, such as DiI, are excellent for tracing membranes. DiI rapidly penetrates along the lipid bilayer and has been used for tracing neurons over distances of many microns. These dyes are not easily soluble in water, so they are typically applied as solids, pastes, or drops of nonaqueous solution. The pastes are made up of the solid dye dispersed in an inert gel. Touching one part of a cell with a paste of DiI on a fine needle will rapidly label the entire plasma membrane. For more general labeling a gene gun can be used (Chapter 12).

As a final point, the cells and the dyes may not tolerate being sealed with nail polish. The solvent has rapid detrimental effects on both viability and fluorescence with vital dyes, even though it is widely used in immunofluorescence of fixed cells. If it is necessary to seal coverslips on slides of living cells, use dental or paraffin wax.

REFERENCES

Burnet, F.M. 1957. A modification of Jerne's theory of antibody production using the concept of clonal selection. *Australian Journal of Science,* 20, 67–69.

Johnson, I., and Spence, M.T.Z., eds. 2011. *Molecular Probes Handbook, A Guide to Fluorescent Probes and Labeling Technologies*, 11th ed. Eugene, OR: Molecular Probes/Invitrogen.

Köhler, G., and Milstein, C. 1975. Continuous cultures of fused cells secreting antibody of predefined specificity. *Nature*, 256, 495–497.

14 Quantitative Fluorescence

FLUORESCENCE INTENSITY MEASUREMENTS

In any project involving quantification of fluorescence, the first consideration is which sort of microscope to use. The confocal microscope is probably the most accurate tool available for measuring intensity within a defined voxel space, but we need to be clear about what this implies. If we want to measure the total fluorescence of an organelle—for example, to determine the DNA content of a nucleus—voxel-based measurement is not really what we want, and widefield fluorescence or flow cytometry is more likely to be a useful tool.

The main problem we cannot escape when making fluorescence measurements is bleaching of the fluorochrome. Bleaching is likely to be at least as significant a factor as any properties of the microscope and measurement system. Second to this is the problem of attenuation with depth in the sample, which is very hard to quantify because it is a mixture of absorption, scattering, and spherical aberration and depends strongly on local properties of the sample. Many problems are best solved by using measures that are independent of intensity, such as ratios (later in this chapter) or fluorescence lifetime (Chapter 15), but for many measurements, such as our example of the DNA content of a nucleus, only intensity measurement will give us the answer (see also Cox, 2011).

LINEARITY CALIBRATION

Photomultiplier tubes (PMTs) are quite reasonably linear through the major part of their range, though they are poor at very low gain (bias voltage)—rarely a problem when measuring fluorescence—and they are often noisy at maximum gain. CCD cameras are very linear, as long as we do not reach the *full well capacity,* the maximum charge that an element of the array can hold. Nevertheless, if we want to make quantitative measurements, it is a good idea to check the linearity of the response.

A simple test specimen is a piece of fluorescent plastic. The filter company Chroma Technology Corp. makes (and often gives away free) fluorescent plastic microscope slides, which are uniform and ideal for the purpose. Alternatively, your local sign writer can probably give you some offcuts; signs commonly use fluorescent (DayGlo) plastic. Adjust the illumination intensity (mercury or laser) using the acousto-optic tunable filter (AOTF) or neutral density (ND) filters (whichever your microscope uses), and plot the image intensity recorded against excitation power. Both AOTFs and NDs are normally accurate, but you can easily check accuracy with a laser power meter, which should be part of the troubleshooting kit in any confocal lab. Take care that the fluorescence does not *saturate* (meaning that all available molecules are in the excited state) at the higher intensities. Mercury lamps change in brightness quite considerably over their lifetime, so you need to recheck at some standard setting each session (preferably both at the beginning and at the end of the session). Lasers also gradually grow dimmer, though this can usually be rectified by realignment, and may also change as they warm up. Both LEDs and metal-halide lamps also fade, but over a much longer time period.

In some respects, an even better test sample is a standard solution of a fluorescent dye, because the molecules diffuse freely, minimizing problems with bleaching. You need a large enough volume to avoid evaporation concentrating the solution. If you have several different concentrations of the fluorochrome solution, you can also check that measured intensity is linear with concentration (it should be, except perhaps at very high concentrations). Whether you use plastic or dye, your standard should, of course, be a similar color to the fluorochrome in your actual experiment, so that you are using the same laser lines and filter sets.

MEASUREMENT

With no illumination you should get a just nonzero level, on average, over the whole image field; otherwise, you may fail to record low levels of fluorescence (Chapter 6). If this is the case, leave the black level or offset at the zero position, which makes life much simpler. If not, adjust the offset accordingly and use this setting for all your experiments. Set your illumination and gain with your calibration sample to give the expected intensity in the image. Start with the brightest sample you want to measure (highest concentration, poststimulus, or whatever) and check that the maximum intensity remains comfortably below saturation. (Otherwise, you will have to set a new calibration value with your standard.) From this point on, the illumination levels and gain controls must not be touched. Write down the values; if you are using a confocal microscope you can probably also save them in an individual method file.

Normally, you will take three measurements:

- Because there will always be background fluorescence in tissues, you must measure an unlabeled sample to quantify this.
- Then measure a labeled negative control.
- Finally, measure your experimental treatments.

The relative differences in the amounts by which the fluorescence in the control and experimental samples exceed the background fluorescence of the unlabeled sample, corrected if necessary for nonlinearity in the detection, are a reasonable measure of the differences in fluorescence resulting from the experimental treatments.

COLOCALIZATION

Often it is useful to know, quantitatively, whether two stains are occurring in the same structure or compartment. The way to measure this is to look at *colocalization*, the extent to which two colors occur in the same pixels. The most common way to represent

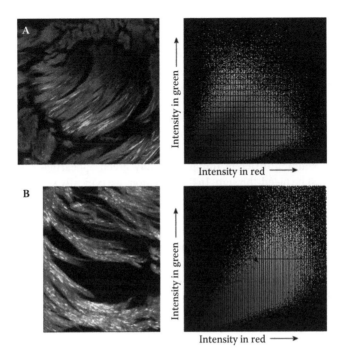

FIGURE 14.1 Images of kangaroo tail tendon and their corresponding colocalization diagrams. Green in each case is the SHG signal from collagen and red is two-photon fluorescence. (A) Fluorescence from Sirius Red staining. There is a substantial degree of colocalization, and it is about equal in both directions. (B) Autofluorescence. There is a higher degree of colocalization for the SHG image. Hardly any green pixels are found above the diagonal, meaning that everything that gives an SH signal also gives a fluorescent signal. The red, however, shows lower colocalization with green; many pixels show fluorescence but not SH. The dark lines in the fluorograms are the "missing" pixel values since both channels have been contrast stretched (Chapter 10).

colocalization is in a *fluorogram*, as shown in Figure 14.1. A fluorogram plots the intensity of a pixel in each of the two channels. A black pixel (no intensity in either) appears at bottom left; a pixel at maximum intensity in both channels appears at top right. Pixels that are totally green spread out along the vertical axis; those that are totally red are along the horizontal axis. Pixels that have some of each color appear somewhere within the picture; along the diagonal if they have equal amounts of both colors. Most confocal microscopes have colocalization software built in, but colocalization is also included in scientific image analysis packages and there are standalone programs as well. Of course, a plug-in exists for ImageJ (http://www.uhnresearch.ca/facilities/wcif/imagej/colour_analysis.htm).

Figure 14.1 gives us an idea of what a fluorogram can tell us. In panel A, although the image seems to show the red and green as not quite matching, there is clearly a lot of colocalization: the fluorogram fans out from the diagonal line. Furthermore, it spreads out equally both sides of the diagonal; there are as many green pixels with not so much red as there are red ones with not so much green. The question being addressed here was whether the second harmonic (SH) signal comes from the same fraction of tendon collagen as that stained by the collagen dye Sirius Red. The answer from the fluorogram is by and large yes but not completely. And the proportion of unstained collagen giving a strong SH signal is similar to that of stained collagen giving a low SH signal. In panel B of Figure 14.1, the SH signal is compared with autofluorescence, and we see almost no green pixels above the diagonal; everything giving an SH signal is also autofluorescent. However, there are a lot of red pixels below the diagonal; quite a lot of autofluorescent regions show little SH signal.

We can also evaluate bleed-through with a fluorogram. If we have two dyes that we know are staining different things, then any green appearing in red pixels (or vice versa) must be bleed-through. In this situation, we typically see a broad group of green pixels running up at an angle in the upper left half, and a similar group of red ones running somewhere above the horizontal axis in the lower right half. Many confocal microscopes can use this information to perform a bleed-through correction.

We can also evaluate colocalization numerically using colocalization coefficients, a method first proposed by Erik Manders (Manders et al., 1993). *Colocalization coefficients* are a variant on a well-known mathematical tool, called Pearson's correlation coefficient. In mathematical notation they look rather daunting:

$$M_1 = \frac{\sum_i Ri, coloc}{\sum_i Ri} \tag{14.1}$$

$$M_2 = \frac{\sum_i Gi, coloc}{\sum_i Gi} \tag{14.2}$$

where $Ri, coloc$ is the intensity in the red of a pixel that also has some green, and Ri is the intensity of any pixel in the red. Likewise, $Gi, coloc$ is the intensity in the green of a pixel that also has some red, and Gi is the intensity of any pixel in the green.

In simpler terms, what M_1 tells us is the fraction of the total red fluorescence that is colocalized with green, and M_2 gives the fraction of the green fluorescence that is colocalized with red. For Figure 14.1B, M_1 is 0.102 and M_2 is 0.913, confirming the impression we got by eyeballing the fluorogram that SH is strongly colocalized with autofluorescence, but autofluorescence has a weaker association with the SH signal.

Certain limitations to colocalization analysis exist. First, bleed-through can give the impression of more colocalization than is present, so we should make sure our channels are separated as well as possible. (One reason for using a second harmonic image as the example here was that in this case bleed-through is almost impossible; the SH signal is monochromatic and shorter in wavelength than any possible two-photon fluorescence.) Second, there is always *dark current*, a small measured signal even where there is no actual fluorescence. To eliminate the effects of dark current, it is normal to set the lowest few gray levels to zero, but this can be an arbitrary choice and, in marginal cases, can skew the results. Third, there is the question of resolution. We cannot expect to tell whether there is really colocalization if two components lie within our minimum resolved distance. The resolution of the objective determines the scale on which we can assess colocalization. To get detail at finer scales, we must turn to FRET, which is covered in Chapter 15.

RATIO IMAGING

Often we want to know something about ions in the cell, perhaps pH or calcium concentration. Many dyes change their properties in response to ion concentration. Some dyes become fluorescent in an appropriate chemical environment. Fluo-3 was one of the first calcium-indicating dyes; it is essentially nonfluorescent unless bound to Ca^{2+} and exhibits a quantum yield at saturating Ca^{2+} of ~0.18. The fluorescence enhancement is of the order of one hundredfold, but nevertheless one might hope for a higher quantum yield, so other dyes have since been developed with higher yields. Figure 14.2 shows a range of visible light–excitable indicator fluorochromes developed by Molecular Probes (Johnson & Spence, 2011). Like fluo-3, these fluorochromes exhibit an increase in fluorescence emission intensity upon binding Ca^{2+} with little shift in wavelength, and can be excited by wavelengths that suit both mercury lamps and typical lasers on confocal microscopes.

Measurements with these dyes (e.g., Figure 9.5) show changes in calcium concentration, pH, and so forth, and these values are quantitative in relative terms. It is very difficult, however, to progress from this to absolute measurements of concentration, because usually we do not know how much dye got into the cell. The best way to obtain absolute measurements of ion concentration is to use *ratiometric imaging*. Here rather than measuring absolute intensity, we measure the ratio between the intensities in two different regions of the spectrum.

This is possible when the wavelength of the emission or excitation peak (or both) changes with the concentration of the ion in question. There are two possible approaches, depending on the characteristics of the particular dye:

- Excitation ratiometry—We can measure the intensity of fluorescence at one wavelength, using two different excitation wavelengths.
- Emission ratiometry—We can use a single excitation and measure the fluorescence at two different wavelengths.

Examples of curves for both emission-ratiometric and excitation-ratiometric dyes are shown in Figure 14.3. In the figure, panels a and b show excitation ratiometry for Ca^{2+} and pH, respectively, where the excitation peak changes substantially, but the emission wavelength remains constant. In contrast, panels c and d show emission ratiometry, where the optimal excitation does not change, but the emission spectrum does. Because all curves for different Ca^{2+} or pH values are distinct, in principle measurements of intensity at any two wavelengths could be used to determine which particular curve we are on and therefore the ionic concentration present. In practice, however, this is not so simple, because the image's signal-to-noise ratio will not

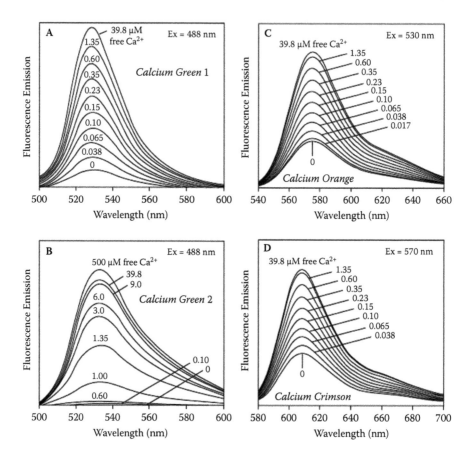

FIGURE 14.2 Calcium dyes developed by Molecular Probes, which show low fluorescence in the absence of calcium but become highly fluorescent when calcium ions are present. (Courtesy of Invitrogen Corp.)

be good enough for any reasonable accuracy unless we choose points where large changes can be expected. As always, for a given integration time, a confocal microscope always gives a worse signal-to-noise ratio than a widefield CCD camera. This disadvantage is offset by the ability to sample in a particular focal plane, so that the measurement is not degraded by interference from cells or free dye in higher or lower planes; only use a confocal microscope for these measurements if you need this ability. Otherwise, you are better off working with a widefield microscope and CCD camera.

The great advantage of ratiometric measurement is that the *ratio* of the intensities at the two wavelengths defines the ion concentration; the absolute intensity is unimportant. It is therefore unaffected by dye concentration, uneven loading, partitioning among cell components, and bleaching. It is, however, affected by background fluorescence, so take care to minimize autofluorescence. Dynamic range is also an issue when dividing one image by another, and 12-bit images are essential

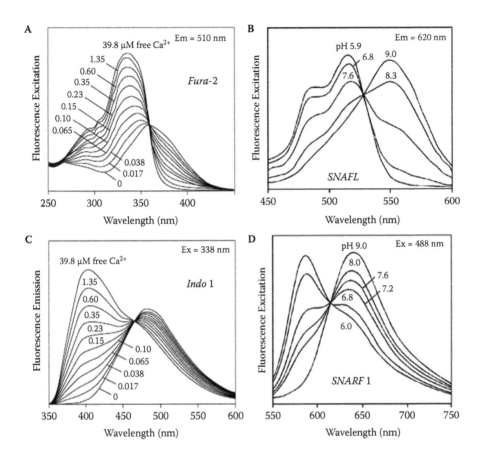

FIGURE 14.3 (A) Excitation spectra at different calcium levels of the ratiometric Ca^{2+} dye Fura-2 when emission is measured at 510 nm. (B) Excitation spectra of the ratiometric pH dye SNAFL when emission is measured at 620 nm. (C) Emission spectra at different calcium levels of the ratiometric Ca^{2+} dye Indo-1 when excited at 338 nm. (D) Emission spectra of the ratiometric pH dye SNARF-1 when excited at 488 nm. (Courtesy of Invitrogen Corp.)

for accurate ratiometry. Take care that even high dye concentrations do not take the intensity at either wavelength to saturation.

In widefield ratiometric imaging, excitation ratiometry is often preferred, because we have an unlimited range of excitation wavelengths available, a filter wheel in front of the light source is easy to implement, and we usually detect in only one channel. For calcium measurement, the dye of choice is usually Fura-2. Images are taken using excitation wavelengths of 340 nm and 380 nm with a 510-nm emission filter, which requires no change of dichroic, and the ratio is taken between the two images.

In the confocal microscope, we have only a finite range of excitation wavelengths and (particularly with spectral detection) an unlimited range of detection wavelengths that can be captured simultaneously, so emission ratio measurements are often the preferred option. However, recent developments, such as the common fitting of four-line argon lasers, have increased the range of wavelengths available on typical confocal microscopes. Similarly, the common use of AOTFs to select wavelengths (so that changes in excitation can be made at high speed) has meant that within the visible range excitation ratiometry is also worth considering. The precision of the laser lines (relative to the rather wide-band filters usually used for widefield emission ratiometry) can give a very accurate measurement if they match the characteristics of the indicator. Fura-2 is not, sadly, a practical proposition in confocal because even ultraviolet (UV) lasers do not have suitable wavelengths; we can get suitable multiphoton wavelengths, but we cannot (yet) switch between them rapidly enough.

Sometimes mixtures of dyes are used for emission ratiometric imaging, such as Fura red and Fluo-3 to measure calcium concentration. Using 488 nm (argon) excitation, ratio imaging can be accomplished by detecting images at 520 nm and 650 nm (therefore routine FITC and rhodamine filter sets could be used). Often one dye is *inert* (it does not change with the ion concentration being measured), whereas the other changes in intensity but not in wavelength. Effectively, therefore, the inert dye acts as a standard. The problem with this approach is that it is difficult to ensure equal loading of the two dyes in all cells of a population. Single dyes are preferable, and the range, now wide, continues to expand.

The fundamental protocol for making an emission-ratiometric measurement is to record images at each wavelength (which can normally be taken simultaneously), subtract the background level (determined previously), and then display the ratio of the two images. Confocal microscope manufacturers normally provide software for doing this automatically, so that the ratio image as well as the two original images are displayed live, which is useful since one knows immediately if the experiment is working. More sophisticated versions can be calibrated so that they will display graphs of ion concentration against time for multiple points over the image. However, if speed is less of a consideration, it is possible to make ratiometric measurements without additional software, using commercial image analysis packages or the freeware ImageJ. If you are using multiline dichroics, take care that they do not interfere with one or another channel; single dichroics are preferable as there is only one excitation wavelength. An example of emission ratiometric measurement in the multiphoton microscope is shown in Figure 14.4.

Excitation ratiometry requires sequential collection of the two images. Assuming that the emission line selection is under software control via an AOTF, fast shutter, or fast filter wheel, the time delay between images will be small (or even nonexistent if excitation can be switched on a line-by-line basis). In general, the best strategy is usually to pick the most effective dye, whether excitation- or emission-based, and build the experiment around it.

In any ratio imaging, accurate calibration is essential when actual numeric output is required. Because both dissociation constants and fluorescent properties may be influenced by the environment inside a cell, this calibration cannot be done in vitro; it must be done on the cell system under investigation. The principle is to clamp or buffer the concentration of the ion in question to a known value or series of values. EGTA can be used to produce defined calcium ion concentrations, and Nigericin is used to calibrate pH. In general, the measured intensity ratios obtained from these

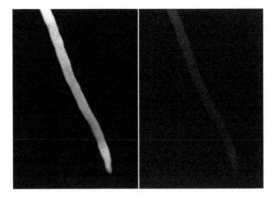

FIGURE 14.4   The two channels of a multiphoton emission ratiometric image of Ca^{2+} in the fish parasite *Saprolegnia ferax*. (Courtesy of Osu Lilje.)

known concentrations can be entered into the microscope's ratiometric software so that the live display shows actual concentrations.

CELL LOADING

Many indicator dyes are not easily taken up by cells, although free acids of Fura-2 and Indo-1 can be loaded into some plant cells at a pH of 4 to 5. One possible loading method is to microinject the potassium or sodium salts of the indicators into cells. Another possibility is to use the acetoxymethyl (AM) esters of these dyes, which passively diffuse across cell membranes. Once inside the cell, these esters are cleaved by intracellular esterases to yield cell-impermeant fluorescent indicators, which do not leak out. A third approach is to use a chemical that induces pinocytosis to persuade the cells to take up the dye in pinocytotic vesicles. By manipulating the tonicity of the medium, these vesicles can then be burst inside the cell, liberating the dye.

MEMBRANE POTENTIAL

Membrane potential is another important parameter in cell physiology. At the cell level, membrane potential often reveals cell–cell communication; within the cell, as in mitochondria, it reveals the activity of individual organelles. Many dyes indicating membrane potential are now available, divided essentially into those dyes that respond slowly, with a substantial shift in fluorescent properties (suitable for physiological studies), and fast-responding dyes, which typically are much more subtle in their response but can monitor rapid signal transduction.

Potentiometric optical probes enable researchers to perform membrane potential measurements in organelles and in cells that are too small to allow the use of microelectrodes. Moreover, in conjunction with imaging techniques, these probes can be used to map variations in membrane potential along neurons and among cell populations with spatial resolution and sampling frequency that are difficult to achieve using microelectrodes.

The plasma membrane of a cell typically has a transmembrane potential of approximately −70 mV (negative inside) as a consequence of K^+, Na^+, and Cl^- concentration gradients that are maintained by active transport processes. Increases and decreases in membrane potential play a central role in many physiological processes, including nerve-impulse propagation, muscle contraction, cell signaling, and ion-channel gating. Potentiometric probes are important for studying these processes, as well as for visualizing mitochondria (which exhibit transmembrane potentials of approximately −150 mV, negative inside matrix).

Probes can be divided into two categories based on their response mechanism:

- *Fast-response probes* (usually styrylpyridinium dyes) operate by means of a change in their electronic structure, and consequently their fluorescence properties, in response to a change in the surrounding electric field. Their optical response is sufficiently fast to detect transient (millisecond) potential changes in excitable cells, including single neurons, cardiac cells, and intact brains. However, the magnitude of their potential-dependent fluorescence change is often small; fast-response probes typically show only a 2% to 10% fluorescence change per 100 mV.
- *Slow-response probes* exhibit potential-dependent changes in their transmembrane distribution that are accompanied by a fluorescence change. The magnitude of their optical responses is usually much larger than that of fast-response probes. Slow-response probes are suitable for detecting changes in average membrane potentials of nonexcitable cells caused by respiratory activity, ion channels, drugs, and other factors, where time is not of the essence.

Potentiometric probes are calibrated by imposing a transmembrane potential using valinomycin in conjunction with K^+ solutions.

FAST-RESPONSE DYES

Styryl potential-sensitive dyes are strongly and directionally affected by electric fields and therefore give both SHG and TPF signals. These dyes respond faster than other potential indicating probes, but show a rather small change in fluorescence. However, the second harmonic signal (Chapter 8) can be much more significant and has now become the preferred technique for measuring fast changes in potential. The major application has inevitably been in neurobiology. The dye FM4-64 has become the probe of choice for measuring action potentials, and Figure 14.5 shows an example.

SLOW-RESPONSE DYES

Derivatives of the carbocyanine dye DiI (Chapter 13) were some of the first potentiometric fluorescent probes. These probes accumulate on the membrane in proportion to its degree of polarization and thence move into the lipid bilayer. Aggregation within the confined quarters of the membrane interior usually results in decreased

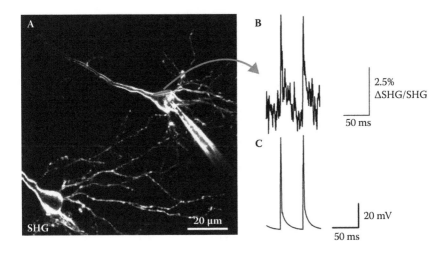

FIGURE 14.5 Fast SHG line scan recording of elicited action potentials (APs) in brain slice. (A) This neuron was patch clamped and filled with FM4-64. Straight red line represents scanned line where elicited action potentials were recorded optically by integrating over the width. (B) SHG recording intensity versus time from 55 averaged scans (1200 lines/s) of the line shown in part A. (C) Average current-clamp trace of elicited APs recorded optically in part B. (From Dombeck, D.A., Sacconi, L., Blanchard-Desce, M., and Webb, W.W., 2005, *Journal of Neurophysiology*, 94, 3628–3636. With permission.)

fluorescence and absorption shifts. Because mitochondria have highly polarized membranes, these dyes tend to stain mitochondrial membranes selectively causing respiratory inhibition and are therefore often relatively toxic to cells.

Rhodamine 123 is widely used as a structural marker for mitochondria and as an indicator of mitochondrial activity. To measure mitochondrial membrane potential, we need to depolarize the plasma membrane by setting the extracellular K^+ concentration close to intracellular values (~137 mM), thereby giving ourselves a reference for the potential difference.

The green fluorescent JC-1 (5,5′,6,6′-tetrachloro-1,1′,3,3′-tetraethylbenzimidazolyl-carbocyanine iodide) exists as a green fluorescent monomer at low concentrations. However, at higher concentrations (aqueous solutions above 0.1 μM) JC-1 forms red fluorescent *J-aggregates*, which exhibit a broad excitation spectrum and a very narrow emission spectrum (Figure 14.6). Uptake of JC-1 into mitochondria varies linearly with applied membrane potential over the range of 30 mV to 180 mV, so its concentration inside the monomer reveals the membrane potential. The amount of the J-aggregate form can therefore be used, by ratiometric comparison with the green fluorescence of the monomer, to measure mitochondrial membrane potential. Figure 14.7 shows an example of this.

FLUORESCENCE RECOVERY AFTER PHOTOBLEACHING

Fluorescence recovery after photobleaching (FRAP) involves bleaching a defined region of a sample with a high laser intensity, and then charting the recovery of

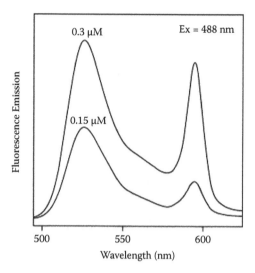

FIGURE 14.6   Spectra of JC-1 at two concentrations, showing that the proportion in the red-fluorescent J-aggregate form increases with concentration.

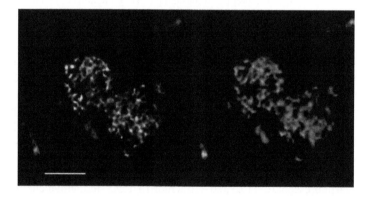

FIGURE 14.7   Mitochondria in living neurons, labeled with JC-1 and imaged with multi-photon excitation. (Left) Two-channel image in which the mitochondria with highest membrane potential appear red. (Right) Ratiometric image of red/green, indicating membrane potential. (From Dedov, V., Roufogalis, B., and Cox, G.C., 2001, *Micron,* 32, 653–660. With permission.)

fluorescence in the bleached region (Figure 14.8). The recovery results from the movement of unbleached fluorophores from the surrounding cytoplasm into the bleached area. FRAP can be used to measure the dynamics of 2D or 3D molecular mobility whether by diffusion, active transport, synthesis, or natural turnover. Sometimes the decrease in fluorescence in the surrounding region is measured instead (or in addition); this is then termed FLIP (*fluorescence loss in photobleaching*) and is shown in the dotted trace in Figure 14.8.

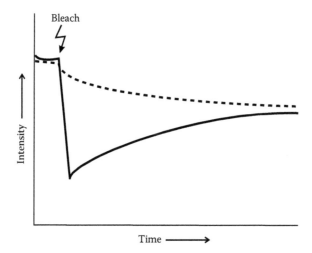

FIGURE 14.8 Progress of a FRAP–FLIP experiment. Before the bleach pulse there will be a little fading. In the bleached area (solid trace) the intensity falls rapidly until the bleach illumination is turned off, then recovers gradually, never reaching its original intensity since some dye has been lost. In the rest of the cytoplasm (dotted trace) the intensity drops gradually as the fluorochrome diffuses into the bleached area; eventually both areas approach the same value.

FRAP has wide applicability in studying cellular dynamics. For example, in studying transport between cell compartments, the fluorochrome in one compartment—the nucleus, for example—can be bleached by high-power irradiation, and the timescale of the recovery of fluorescence in that compartment can show whether the mechanism is active transport or diffusion. The final intensity level after recovery shows whether any de novo synthesis is involved.

Confocal microscopes are usually used for FRAP experiments because they allow for pixel-by-pixel control, especially if the microscope has an AOTF to control the laser input (Chapter 5). If the microscope does not have an AOTF, we are restricted to bleaching rectangles or lines, but that can be very effective (Figure 14.9), not least because it simplifies numerical calculations. When the excitation is controlled with an AOTF, it is possible to irradiate a precise area by turning on the excitation as the spot passes the desired area. Thus, a specific cell, or cell compartment, can be bleached quite precisely. Some microscopes are now equipped with a second scanner specifically for photobleaching, photoactivation, and photoconversion, so that the bleaching or converting scan can be set up independently of the imaging scan. This is essential on line scan and spinning disk systems but is also useful on conventional single-point confocals.

FRAP has contributed to much of our knowledge of protein dynamics in the cell. It was used to help define the fluid mosaic model of cell membranes. One particularly effective use has been in studying the dynamics of tubulin in microtubules. Fluorescently tagged tubulin is introduced into the cells by microinjection, and sections of microtubule arrays are then bleached. As new tubulin is incorporated into the

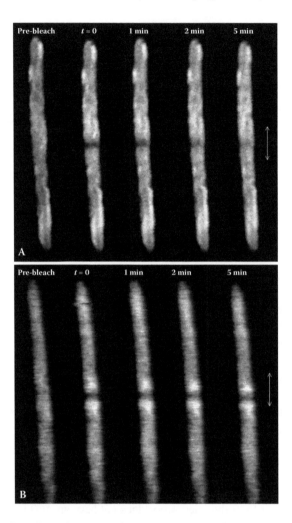

FIGURE 14.9 Cyanobacterium *Dactylococcopsis salina*. (A) Fluorescence excited with a 633 nm laser reveals the phycobilisomes. The bleached area gradually spreads and the bleach becomes shallower, indicating diffusion of phycobilisomes on the membrane surface. (B) With a 442 nm laser the fluorescence observed is from photosystem II complexes in the thylakoid membranes. In contrast to the result for phycobilisomes there is no detectable diffusion of photosystem II. (From Mullineaux, C.W., Tobin, M.J., and Jones, G.R., 1997, *Nature*, 390, 421–424. With permission.)

microtubules, they gradually become fluorescent again. This enables us to study the kinetics of tubulin turnover, showing, for example, that the dynamic microtubules of the spindle in cell division turn over tubulin twice as fast as cytoskeletal microtubules in interphase cells. FRAP can also be used to determine whether a protein can move within a membrane (high percent recovery with a fast mobility) or whether the protein is tethered to other structural components of the cell (low percent recovery with a slow mobility), as shown in Figure 14.9.

FIGURE 14.10 HeLa cells stably expressing Kaede. (A) Imaged with blue light. (B) Irradiating specific areas with violet light (405 nm laser) changes the fluorescence from green to red; exchange of the converted protein is rapid through the cytoplasm but interchange between nucleus and cytoplasm is slow. (Courtesy of Atsushi Miyawaki.)

The major problem with FRAP is that the laser intensities involved in bleaching may damage the cell. Some carefully controlled studies have demonstrated that there was no damage (Hepler and Hush, 1995); others have shown clear signs that damage is occurring. Thus, although FRAP without damage is possible, the risk of damage is real. This is where color-changing proteins, such as Kaede, come into their own. Kaede (Chapter 12) is an expressible fluorescent protein that normally fluoresces green, but after irradiation with UV or deep violet light turns red. The light intensity required to do this is far less than that involved in photobleaching, so we can use the color change to study cell dynamics (Figure 14.10) with equivalent methodology to conventional FRAP but with less risk of damage and therefore higher confidence in the results. Photoactivatable proteins such as Kindle and Dronpa (Chapter 12) can be used in the same way. With the recent development of photoswitchable probes extending into the far-red region (Subach et al., 2011) it has even become possible to consider using two different proteins in one dynamic experiment, which has some exciting possibilities.

REFERENCES

Cox, G.C. In press. Measurement in the confocal microscope. In *Confocal Microscopy, Methods and Protocols,* 2nd ed., S. Paddock, ed. Totowa, NJ: Humana Press.

Dedov, V., Roufogalis, B., and Cox, G.C. 2001. Visualisation of mitochondria in living neurons with single and two-photon fluorescence laser microscopy. *Micron,* 32, 653–660.

Dombeck, D.A., Sacconi, L., Blanchard-Desce, M., and Webb, W.W. 2005. Optical recording of fast neuronal membrane potential transients in acute mammalian brain slices by second-harmonic generation microscopy. *Journal of Neurophysiology,* 94, 3628–3636.

Hepler, P.K., and Hush, J.M. 1996. Behavior of microtubules in living plant cells. *Plant Physiology,* 112, 455–461.

Johnson, I., and Spence, M.T.Z., eds. 2011. *Molecular Probes Handbook, A Guide to Fluorescent Probes and Labeling Technologies,* 11th ed. Eugene, OR: Molecular Probes/Invitrogen.

Manders, E.M.M., Verbeek, F.J., and Aten, J.A. 1993. Measurement of co-localization of object in dual-colour confocal images. *Journal of Microscopy*, 169, 375–382.

Mullineaux, C.W., Tobin, M.J., and Jones, G.R. 1997. Mobility of photosynthetic complexes in thylakoid membranes. *Nature*, 390, 421–424.

Subach, O.M., Patterson, G.H., Ting, L.-M., Wang, Y., Condeelis, J.M., and Verkhusha, V.V. 2011. A photoswitchable orange-to-far-red fluorescent protein, PSmOrange. *Nature Methods,* 8, 771–777.

15 Advanced Fluorescence Techniques
FLIM, FRET, and FCS

FLUORESCENCE LIFETIME

When a molecule absorbs light, we do not get fluorescence immediately. As we saw in Chapter 3, the electron must lose some energy before it reaches the lowest vibrational level of the S_1 state (Figure 15.1). Only from there can fluorescence take place (Kasha's law). We will therefore see an initial delay at excitation, and then (in the absence of outside events such as resonant energy transfer [FRET] or chemical reactions) decay will follow first-order kinetics—an exponential decay curve—because only one chemical species is involved (Figure 15.2).

The formula describing first-order kinetics is

$$\frac{I}{I_0} = e^{-kt}$$

where I is the intensity at time t, I_0 is the initial intensity, and k is the fluorescence decay rate (e is the base of natural logarithms and has a value approximately 2.72).

The fluorescence lifetime τ is defined as

$$\tau = \frac{1}{k}$$

or, in other words, the time taken to decay to $1/e$ of the original intensity.

Because fluorochromes differ in lifetime (Table 15.1) we can, in principle, distinguish among them by lifetime, even if their emission spectra are identical. Furthermore, we can expect the lifetime to change if anything happens in the chemical environment around the molecule that affects fluorescence. If any other de-excitation mechanism starts to compete with fluorescence, we no longer have first-order kinetics because two or more processes are involved. Instead, we have a curve following second- or third-order kinetics, and there is a marked reduction in lifetime.

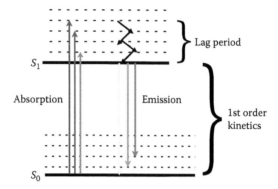

FIGURE 15.1 Jablonski diagram showing the time factors involved in fluorescence emission. No fluorescence can be emitted until electrons have reached the lowest vibrational level of S_1. From then on, decay follows first-order kinetics.

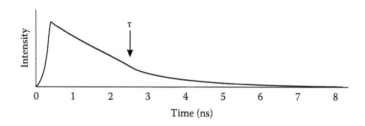

FIGURE 15.2 Typical lifetime curve for a fluorescent pigment. There is a rapid rise followed by a slow, exponential decay. The lifetime τ is taken as the time at which it has decayed to $1/e$ (~0.37) of the peak intensity.

TABLE 15.1
Lifetimes of Some Common Fluorochromes

Fluorochrome	Lifetime (τ)
Eosin	1.1 ns
Lissamine rhodamine	2.13 ns
GFP	2.45 ns
YFP	2.9 ns
Fluorescein	3.25 ns
Texas red	3.41 ns
Lucifer yellow	7.81 ns

PRACTICAL LIFETIME MICROSCOPY (FLIM)

If fluorescence lifetime is to be useful to us, we must be able to measure it. Chemists do this in cuvettes with a fast flashlamp, a photodiode, and an oscilloscope, but in cell biology we need to use a microscope and measure the lifetime at each point in the image—*fluorescence lifetime imaging* (FLIM). How can we achieve this in practice? Two approaches have been used (Chang et al., 2007):

- Frequency domain—Measuring how far fluorescence is out of phase with a modulated excitation
- Time domain—Measuring the rise and decay of fluorescence directly after a pulse of light

FREQUENCY DOMAIN

In the frequency domain approach, the excitation light is modulated either with an on–off modulation (square wave) or a sine wave. The simplest (and cheapest) way to do this with modern technology is to use a high-intensity LED as the light source, because this can be modulated directly. Older designs used a mercury lamp, or even a laser, modulated by some form of fast shutter, typically an electro-optic modulator (EOM). The longer the lifetime, the further out of phase with the excitation modulation the detected signal will be.

Alternatively, we can measure the extent of demodulation of the fluorescent signal relative to the excitation. Using a lock-in amplifier to control the phase of signal detection in relation to the modulation of the incoming beam, we can, for example, find that the FITC signal has a smaller phase shift than lucifer yellow (see Table 15.1). Confocal microscopes working on this basis have been built (Carlsson and Liljeborg, 1998), but the commercial implementations of frequency domain lifetime microscopy are widefield systems with LED excitation (Figure 15.3). Essentially, with such a system, the user adjusts the phase until the particular feature of interest is brightest. From the phase shift he or she can estimate the fluorescence lifetime. Several images, at different phase shifts, may be needed to cover all the relevant regions of interest.

Frequency domain systems are fast, photon-efficient, and relatively inexpensive. Their main disadvantage is that they do not give an actual lifetime profile. It is an indirect measurement so its accuracy depends on calibration, and the images, particularly if complex, may need careful interpretation.

TIME DOMAIN

Time domain methods are most easily applied in confocal microscopy where we deal with each point (pixel) individually. A pulsed laser is required, and the alternatives normally used are either a pulsed diode laser at 405 nm, 440 nm, or 470 nm (single photon excitation) or the natural pulses of a titanium–sapphire laser, which implies multiphoton excitation (Chapter 8). The two best-known commercial systems work by photon counting but differ considerably in their design philosophy.

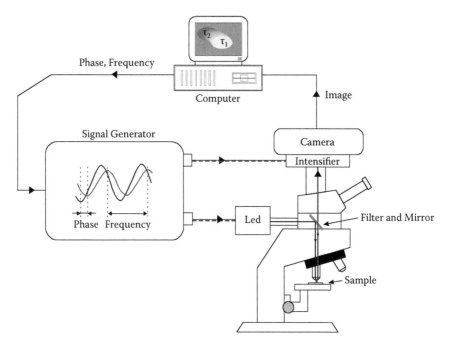

FIGURE 15.3 Layout of the Lambert Systems LIFA frequency domain widefield lifetime imaging system. (Courtesy of Lambert Systems.)

Time-correlated single-photon counting (TCSPC; Becker, 2010) is the basis of a commercial system made by Becker and Hickl, which can be fitted as an accessory to most confocal microscopes and used with pulsed visible lasers or multiphoton excitation. The typical repetition rate for a Ti-S laser is 80 Mhz, which means that the time between pulses is 12.5 ns. There are about 200 of these pulses at each pixel as the beam scans the sample. From each pulse, only one photon is recorded and put in a bin according to its time of arrival. Eventually, the contents of the bins record a full lifetime histogram (Figure 15.4).

The TCSPC system gives a very high resolution spectrum, with up to 256 bins, so very accurate curve fitting can be carried out. It is possible to distinguish between single exponential and multiexponential decay (and extract up to three components in the multiexponential case). However, this approach requires a lot of photons, and image acquisition times can be long, though improvements in detectors have improved efficiency in the latest models.

A different approach in the *time-gated* technique used is the Nikon (Europe) LIMO system, an accessory for the Nikon C1 confocal microscope. This system was developed by Hans Gerritsen at the University of Utrecht (Gerritsen et al., 2002). After each pulse, we collect photons for 2 ns into one window, then for 2 ns into the next, and so on, collecting four different channels at each pixel (Figure 15.5). Curve-fitting to the four intensities stored for each pixel then enables the lifetime to be calculated. The time-gating technique is a high-speed, photon-efficient system, suitable for use on living cells, but it will not give the spectral resolution of the TCSPC approach.

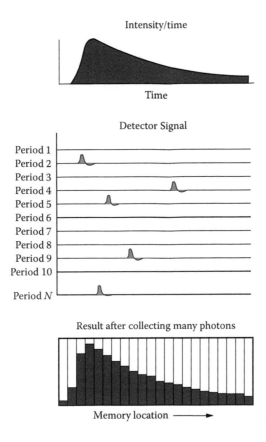

FIGURE 15.4 Acquisition of a spectrum at any one pixel in the TCSPC system. With each pulse, a single photon is counted and stored in a bin, depending on its time of arrival. After many pulses have arrived, a complete spectrum is built up. This process is repeated for each pixel in the image. (Courtesy of Becker & Hickl GmbH.)

An example of the sort of information we can obtain from lifetime imaging is shown in Figure 15.5 and Figure 15.6. Chlorophyll functions in the living cell by transferring its energy to photosynthesis, which is a competing pathway for de-excitation; energy that goes to photosynthesis is not released as fluorescence. The lifetime is therefore short (Figure 15.5). When the cells are no longer healthy, photosynthesis decreases, there is less de-excitation from that source, and the lifetime increases (Figure 15.6).

Other chemical changes that can affect fluorescence lifetime include such variables as pH and calcium concentration, with appropriate indicator dyes. Instead of using changes in color or intensity (which must be measured ratiometrically if they are to be quantitative), we can measure lifetime and get quantitative measurements directly.

Many indicators that vary in fluorescent intensity with changes in ion concentration (e.g., Calcium Green) also show a lifetime change. Because their wavelength response does not change, they cannot be used ratiometrically; in conventional microscopy these dyes can be used only for qualitative imaging (Chapter 14). However, lifetime

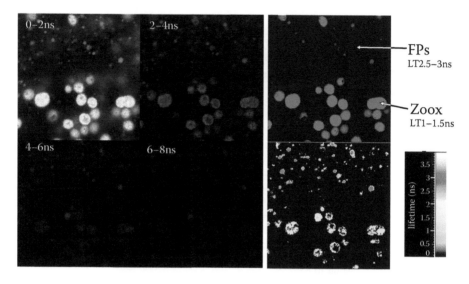

FIGURE 15.5 Time gated lifetime imaging with the LIMO system, using single photon excitation from a pulsed 405 nm diode laser. (Left) The four time windows. The sample is a squash of living coral tissue containing symbiotic dinoflagellate algae (zooxanthellae) and fluorescent pigments (FPs). (Right, upper) Intensity image from the sum of all four time channels; (lower) lifetime image. The chlorophyll autofluorescence shows a much shorter lifetime (1–1.5 ns) than the fluorescent pigments (2.5–3 ns). (From Cox, G.C., Matz, M., and Salih, A., 2007, *Microscopy Research & Technique*, 70, 243–251. With permission.)

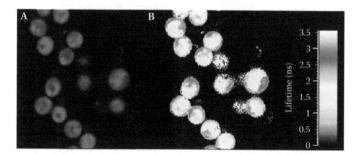

FIGURE 15.6 Zooxanthellae from a similar coral squash but after storage overnight. The morphology of the algae looks unchanged (A) but the lifetime of the chlorophyll fluorescence has increased to 2–2.5 ns (B), indicating that the pigment is no longer coupled to photosynthesis. (From Cox, G.C., Matz, M., and Salih, A., 2007, *Microscopy Research & Technique*, 70, 243–251. With permission.)

is largely independent of dye concentration, at least at likely concentrations in living cells, so these dyes can be used quantitatively in a FLIM system. Calibration protocols are essentially the same as for ratiometric dyes (Chapter 14). This makes a wide range of simple dyes available for quantitative work, using only one excitation wavelength and one detection channel. What is particularly convenient for time-course studies is that FLIM measurements are quite robust to bleaching of the indicator.

FLUORESCENCE RESONANT ENERGY TRANSFER (FRET)

Fluorescence (or *Förster*) *resonant energy transfer* (FRET) is an interaction between the electronic excited states of two dye molecules and was first studied by Theodor Förster (1948). Excitation is transferred from a donor molecule to an acceptor molecule without emission of a photon. The emission spectrum of the donor must overlap the excitation spectrum of the acceptor; the extent of this overlap determines the FRET efficiency, $J\lambda$. Also, donor and acceptor transition dipole orientations must be approximately parallel. The electronic interactions involved are shown as a Jablonski diagram in Figure 15.7.

FRET depends on the inverse sixth power of the intermolecular separation, so it falls off extremely rapidly with distance. This makes it useful over distances comparable with the dimensions of biological macromolecules, typically 2 nm to 50 nm. When FRET occurs, the fluorescence of the donor molecule is quenched and only the acceptor fluoresces, but the excitation maximum is that of the donor. FRET gives us a powerful tool for measuring the proximity of molecular species at a level hugely beyond the resolution of any microscope.

The *Förster radius* R_0 is the distance at which energy transfer is 50% efficient (in other words, where 50% of excited donors are deactivated by FRET). Table 15.2 gives some examples.

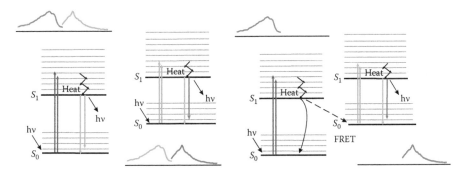

FIGURE 15.7 (Left) When two suitable fluorochromes are separated by more than the Förster radius, most molecules will absorb and emit independently. (Right) When they are closer than this, energy absorbed by the donor will pass to the acceptor by FRET.

TABLE 15.2
Some Typical Values for the Förster Radius

Donor	Acceptor	Ro (nm)
Fluorescein	Tetramethyl-rhodamine	5.5
IAEDANS	Fluorescein	4.6
EDANS	DABCYL	3.3
Fluorescein	QSY-7 dye	6.1
CY3	CY5	5

If we know the Förster radius we can calculate the molecular separation from the FRET efficiency:

$$E = \frac{1}{1 + \left(r/R_0\right)^6}$$

WHY USE FRET?

A common use of FRET is to look for molecular interactions in a cell. We can examine such things as dimer/polymer formation or enzyme–substrate binding within a cell by attaching suitable labels to the two components and looking for FRET. The GFP family of proteins is a common choice for this type of experiment, because these proteins offer a range of wavelengths and can be expressed in cells (Figure 15.8). CFP–YFP are the most popular choice, because both confocal and widefield microscopes have suitable lines available in the violet-indigo region. Both the 457-nm argon line and 405-nm diode lasers effectively excite CFP.

In some situations, we wish to know whether a particular target is located in a particular cellular compartment, and for this we might want to use immunolabeling or organelle/compartment probes. Figure 15.9 shows examples of common fluorochromes that make effective FRET pairs.

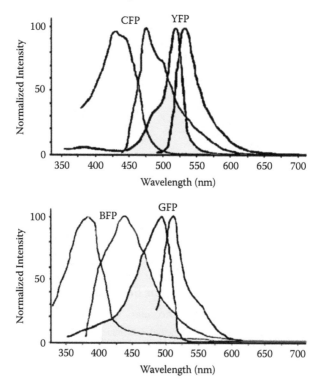

FIGURE 15.8 Examples of GFP variants used as FRET pairs. The colored area indicates the spectral overlap and hence the FRET efficiency Jλ.

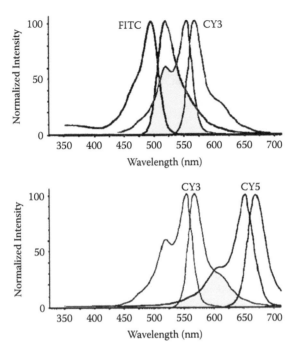

FIGURE 15.9 Non-GFP FRET pairs and examples of dyes that could be useful for immunolabeling FRET experiments.

Donor and acceptor can even be in one molecule, as in the cameleons, which consist of two GFP derivatives joined by Ca^{2+}-sensitive contractile link (Brasselet et al., 2000). These *expressible indicators* are based on a CFP separated from a YFP by the Ca^{2+}-binding protein calmodulin (CaM) and a calmodulin-binding peptide (M13) (Figure 15.10). This allows ratiometric measurement of calcium ion concentrations in cells by FRET. If Ca^{2+} ions are bound, CaM wraps around M13 (lower diagram), and the construct forms a more compact shape leading to a higher efficiency of excitation transfer from the donor CFP to the acceptor YFP. Emission switches from 480 nm to 530 nm. This gives us a ratiometric indicator we can express in the cell rather than having to load it.

IDENTIFYING AND QUANTIFYING FRET

We can identify and measure the extent of FRET in several ways, each of which has its own advantages and disadvantages.

Increase in Brightness of Acceptor Emission

This is the most obvious criterion and often the one we must use, because it almost always gives the best signal-to-noise ratio. Also, it does not need any extra hardware on the microscope. However, this approach presents many problems for accurate measurement. We must make two important corrections before we can evaluate the degree of FRET observed in an image:

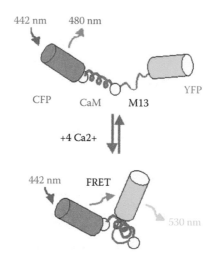

FIGURE 15.10 Diagrammatic view of a cameleon. (Brasselet, S., Peterman, E.J., Miyawaki, A., and Moerner, W.E., 2000, *Journal of Physical Chemistry B*, 104, 3676–3682.)

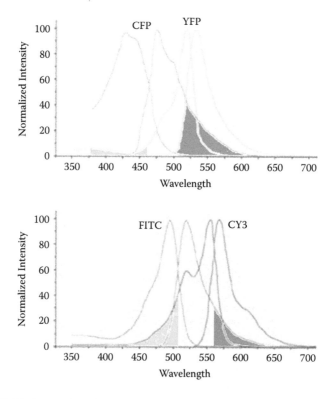

FIGURE 15.11 Examples of excitation cross talk (blue) and emission bleed-through (orange). In the CFP–YFP combination there is little excitation cross talk, whereas the FITC–CY3 pair has a lot. Both show substantial emission bleed-through.

- Excitation crosstalk—Direct excitation of the acceptor by the donor excitation wavelength
- Bleed-through—Emission from the donor at the wavelengths being measured for the acceptor

For efficient FRET the two dyes must be spectrally fairly close, so both these corrections are often significant (Figure 15.11). In fact, balancing the two properties will always be tricky, since maximizing efficiency also maximizes these problems. We therefore need to make corrections for either cross talk or bleed-through, usually both, in every experiment. Figure 15.12 shows a worked example.

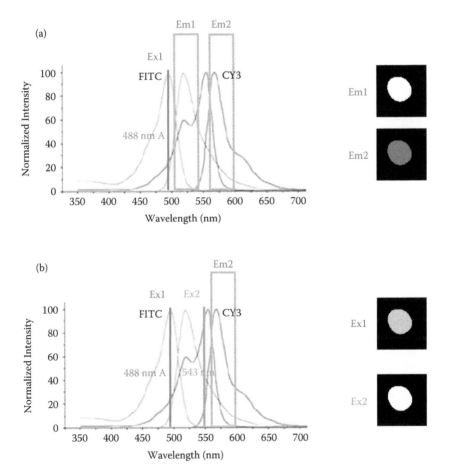

FIGURE 15.12 Correction of intensity-based FRET, a worked example. (a) Sample with donor alone: measure fluorescence emission at both detection wavelengths (Em1, Em2) using donor excitation wavelength. Calculate the fraction of Em1 seen in Em2. In this example we find it is 10%, or 0.1. (b) Sample with acceptor alone: measure emission from both excitation wavelengths. Calculate the fraction of Ex2 seen in Ex1. In our worked example we find it is 20%, or 0.2.

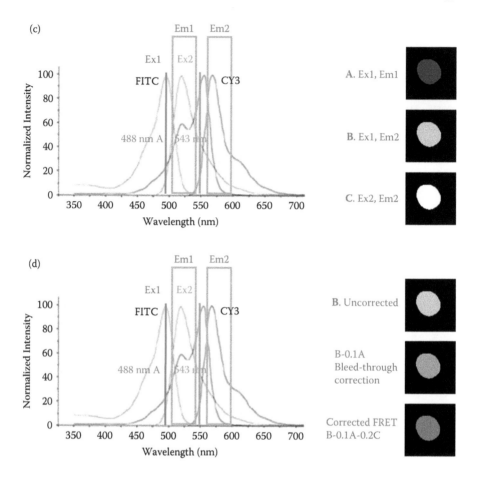

FIGURE 15.12 (continued). (c) Sample with both: measure emission at both detection wavelengths from excitation 1 and at the acceptor detection wavelength from excitation 2. Image A is the donor, depleted by FRET; image B is the *uncorrected* FRET image, and image C is the normal (non-FRET) acceptor image. (d) The top image is the uncorrected image B from (c). Calculate 10% of the intensity of image A—this is the bleed through correction from (a)—subtract that to get the center image. Calculate 20% of the intensity of image C—this is the cross-talk correction from (b)—and subtract that also. This gives the bottom image which is the corrected FRET image, free of both donor fluorescence bleed-through and donor excitation cross-talk.

Fortunately, software ("pFRET," an ImageJ plug-in developed in Ammasi Periasami's Virginia lab) is now available to take the drudgery out of the calculation. However, one still must collect seven different images to make a single measurement (Elangovan et al., 2003). Figure 15.13 shows an actual FRET experiment—oligomerization of membrane glutamate receptors—quantified in this way with the pFRET plug-ins.

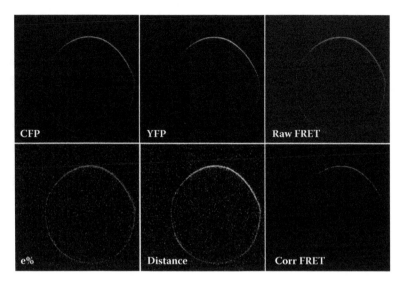

FIGURE 15.13 Oligomerization of membrane glutamate receptors. *Xenopus* oocyte, expressing both receptor tagged with CFP and the same receptor tagged with YFP. If they form oligomers FRET will occur. The raw image suggests that FRET is occurring, and the corrected image (made with pFRET) confirms this. The software also generates maps of FRET efficiency (e%) and from that a distance map. (Courtesy of Anne Mitrovic.)

Quenching of Emission from the Donor

In some experiments, introduction of the acceptor will cause an immediate loss in fluorescence emission from the donor. The decrease in donor emission can often be a more accurate indicator than the increase in acceptor emission because it is free from emission bleed-through and excitation cross-talk problems. As such, this is useful only in dynamic experiments, where we start out either without the donor present or without FRET.

However, we can use the same principle in static experiments by doing the procedure in reverse: bleaching the FRET acceptor and looking at the resulting increase in donor fluorescence. With most FRET pairs, the optimum excitation wavelength for the acceptor does not excite the donor to any significant extent. Therefore, an extended exposure to the acceptor excitation destroys the acceptor and eliminates FRET. The method is to measure donor emission, and then bleach the acceptor and measure donor emission again (Figure 15.14). The increase in donor emission is a direct measure of the amount of FRET that was originally present. An example is shown in Figure 15.15. This approach is simple and widely used, but it does have the problem that other phototoxic effects could occur as a side effect of the bleaching, influencing the result.

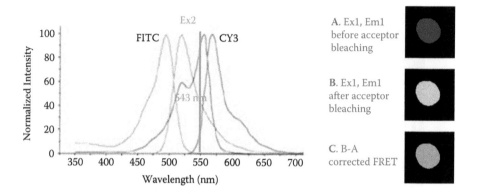

FIGURE 15.14 FRET correction by acceptor bleaching. 543 nm or 561 nm excitation will excite, and therefore ultimately bleach, Cy3 but does not excite FITC at all. The increase in donor fluorescence (image B, postbleach, minus image A, prebleach) shows how much FRET there was before the acceptor was destroyed (image C).

Lifetime of Donor Emission

FRET provides an alternative pathway for de-excitation, which means that it reduces the lifetime of the acceptor. This provides an impartial quantitation technique that does not depend on a large number of controls. In the presence of FRET, the fluorescence lifetime, τ, is reduced; the decay curve is also no longer a single exponential because two chemical processes are involved. This makes FLIM an excellent tool for quantifying FRET, and in many cases it will be the method of choice. Figure 15.16 shows an example.

The great advantage of this approach to FRET is that quantification is simple, as long as only two FRET participants are involved. The FRET efficiency, E, is given by

$$E = 1 - \frac{\tau_{da}}{\tau_d}$$

where τ_d is the lifetime of the donor on its own and τ_{da} is the lifetime of the donor in the presence of the acceptor.

The main limitation to FLIM as a measure of FRET is that the signal must be strong, because signal-to-noise ratio is inevitably limited by the need to sort the number of photons both by wavelength and by time. However, the measurements are clear and unambiguous. In particular, acceptor intensity-based approaches to FRET cannot distinguish between FRET and radiative transfer, where emission from the donor excites fluorescence from the acceptor. Radiative transfer depends on the inverse square of the distance between the two molecules and therefore is much less sensitive to distance than FRET.

This approach naturally depends on the availability of a lifetime imaging microscope, but these systems are now much more affordable than when the first examples reached the market. The avoidance of the need to collect so many controls, and of the assumptions that have to be made about these controls, makes it seem that

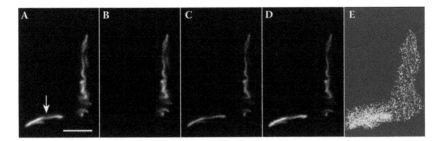

FIGURE 15.15 Labeling of acetylcholine receptor (AChR) using TRITC-alpha-bungarotoxin (donor) and Alexa647-alpha-bungarotoxin (acceptor) in a FRET assay to determine the changes in distance between receptors caused by denervation. The toxin binds to the AChR at the neuromuscular junction (mouse tibialis anterior in this series). (A) Acceptor fluorescence, excited at the acceptor wavelength (633 nm). The endplate (arrowed) in the lower part of the image is then selectively bleached at the acceptor excitation wavelength (633 nm). (B) Acceptor fluorescence after bleaching of the region containing the endplate. (C) Donor fluorescence image (at the donor excitation wavelength 543) before acceptor bleaching. (D) Corresponding donor fluorescence images after acceptor bleaching. The increase in donor fluorescence is evident. (E) Calculated FRET efficiency from panels C and D, in a "spectral" palette where the highest value represents 50% efficiency. Scale bar = 5 μm. (Images courtesy Dr. Othon Gervasio.)

lifetime imaging is likely to become more often the method of choice for assessing and quantifying FRET.

Protection from Bleaching of Donor

Any donor molecules associated in a FRET pair have an additional pathway for de-excitation in operation, making them much less susceptible to bleaching than the free donor. The protection from bleaching relates directly to the reduction in τ (previous section) so measuring this protection is essentially equivalent to measuring lifetime (Jovin & Arndt-Jovin, 1989). A series of images is taken while the sample bleaches, to which a curve can be fitted and the bleaching lifetime, τbl, calculated (τbl being the time taken to bleach to 1/e of the starting value). τbl is inversely related to the fluorescence lifetime (shorter lifetime means slower bleaching, longer τbl) and can be used to calculate FRET efficiency, E, in the same way as if fluorescence lifetime had been measured, by simply inverting the equation:

$$E = 1 - \frac{\tau bl_d}{\tau bl_{da}}$$

Fitting a curve to the bleaching progress is better than simply measuring the decay to 1/e of the original value, because some bleaching is likely to have taken place during focusing and location of the region of interest. The control to establish the bleaching rate of the donor in the absence of FRET is normally a separate sample (unless you are lucky enough to have the donor present in two populations: one involved in FRET and the other not). Because many factors can influence bleaching rates, you should take and average several measurements to establish τbl_d accurately, and you should check laser intensity at the beginning and end of each experiment.

In principle, this method is equivalent to lifetime imaging for measuring FRET but does not require any special hardware. However, it is very slow, and the system must be static because nothing must move during the measurement of the bleaching curves. It could never be used on living cells. The acceptor also tends to bleach, reducing FRET and skewing the result. This technique has not become popular.

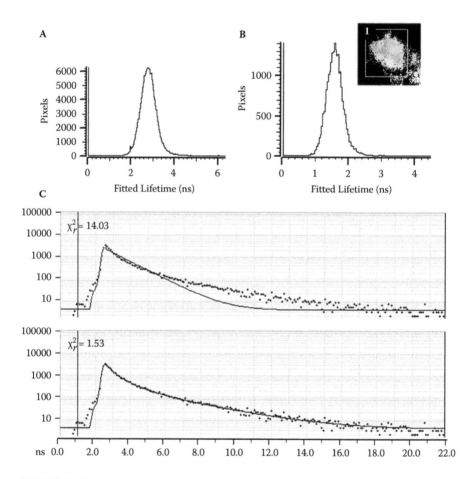

FIGURE 15.16 Reduction in donor lifetime by FRET. (A) Lifetime histogram of pure pigment (amil504) from the coral *Acropora millepora*. Modal lifetime is 2.9 ns. (B) Lifetime histogram and image of pigment granules in the coral, which contain three pigments. Modal lifetime is 1.56 ns, showing that FRET is taking place (Nikon LIMO system). (C) Actual decay curve from the coral pigment granule (points); a single exponential decay (solid line, upper trace) is a very poor match, while a triple exponential decay fits very well (lower trace). Excitation 405 nm (Becker & Hickl system). (From Cox, G.C., Matz, M., and Salih, A., 2007, *Microscopy Research & Technique*, 70, 243–251. With permission.)

FLUORESCENCE CORRELATION SPECTROSCOPY (FCS)

In its original (and still widely used) form (Magde et al., 1972) *fluorescence correlation spectroscopy* (FCS) is not an imaging technique (so arguably does not belong in this book), but it is normally carried out on a confocal microscope (reviewed by Hess et al., 2002).

If we stop the beam scanning to look at just one spot in the confocal microscope, using a high numerical aperture objective, the excited volume is very small indeed. If we collect rapidly enough, the signal at the detector fluctuates wildly. Some of this fluctuation will be true *shot noise*, the inherent randomness of photon emission and detection (Chapter 6). However, another contribution to the fluctuation comes from molecules moving (by diffusion of otherwise) in or out of the tiny volume being sampled. The bare trace from the PMT looks chaotic (Figure 15.17), and at first glance it would seem impossible to get any useful information from it.

The secret to extracting information from the chaos is conceptually not difficult but does require significant computation. The idea is to carry out an *autocorrelation* by going through the entire trace and looking at the likelihood of fluorescence present at any one time, t, still being there at some later time, $t + \tau$. We do this check (on the entire data stream) for a range of values of τ from microseconds to seconds. Doing so gives us some idea of whether fluorochrome molecules are moving rapidly (short duration, low degree of autocorrelation) or slowly (longer duration, high degree of autocorrelation). We plot this probability as a curve, which is much simpler to interpret than the raw data. Hypothetical examples are shown in Figure 15.18 (in which each trace represents a separate experiment). A free dye molecule has a short residence time, with very few found after 0.1 ms (left curve), whereas a dye molecule bound to a substantial protein molecule moves slowly, typically still present up to

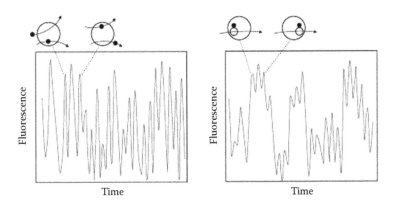

FIGURE 15.17 Two hypothetical traces of raw FCS data. Free fluorescent molecules move in and out of the detection volume (open circle) and are detected as a series of short, randomized fluorescence bursts (left panel). Macromolecule-bound ligands are less mobile, producing a more slowly fluctuating (i.e., more highly autocorrelated) time-dependent fluorescence pattern (right panel). (Courtesy of Invitrogen Corp.)

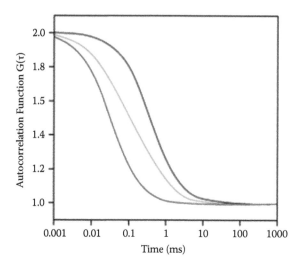

FIGURE 15.18 Simulated FCS autocorrelation functions representing a mobile fluoro-chrome (blue curve), macromolecule-bound fluorochrome (red curve), and a 1:1 mixture of free and bound fluorochromes (green curve). (Courtesy of Invitrogen Corp.)

1 ms (right curve). A mixture of the two samples (middle curve) lies in between as expected, but the slope of the curve is different showing that two species are present.

Evidently, interesting data can be extracted from the apparent random flickerings of the fluorescence signal. What can we use it for? The mobility of membrane components is one possibility, as is protein–protein interaction (is the labeled component free or bound?). We should be able to distinguish between diffusion and active transport, and investigate the free or incorporated status of nucleic acids.

In systems where two molecules of roughly equal mass interact, the logarithmic nature of the FCS plot (Figure 15.18) means that the difference between free and bound molecules is below the detection threshold. To handle this situation, a further refinement of the technique is to use two detection channels and carry out a cross-correlation analysis. In this way, we can clearly separate the signals from the free and bound components. The problem here is that few lenses are totally achromatic; even if they are, the wavelength difference means that the sampled volume is different in each channel. One solution is to use two-photon excitation, because in this case the excitation volume is determined only by excitation, and chromatic and wavelength effects are eliminated (Chapter 8). This inevitably adds to the expense, but the other benefits of two-photon microscopy are sufficiently substantial that the hardware is not rare. At the time of this writing, however, serious biological applications of fluorescence cross-correlation spectroscopy remain rare.

RASTER IMAGE CORRELATION SPECTROSCOPY

Information from one point is all well and good, but wouldn't it be better if we could get this information over an entire image? We could then see where things were bound and where they were mobile. This is the basis of *raster image correlation spectroscopy* (RICS; Digman and Gratton, 2007).

The idea is to hugely oversample, collecting up to 10 points within the Airy disk (4–5 times Nyquist sampling). Since this is well below the resolution limit, things should not change much from point to point unless fluorescent molecules are moving in or out of the excited volume. The movement from point to point on a line scan will pick up very rapid movement, whereas the variations from one line scan to the next will detect slower movement, and the difference from frame to frame even slower changes. Of course there is a lot of complex math involved to sort out the fluctuations from the natural change from point to point, but in the end we get a lot of useful information. How we put this back together depends on our sample. For example, in a living cell we may use the longer time scales to map natural cell movement and recompute the faster changes to compensate for that. The key thing is that we can separate static items from moving ones and get an idea of the mechanisms of the different types of movement.

RICS is now a commercially available product, licensed from the Gratton lab by Zeiss (who have always been big in FCS). So far it is in its early days, but this could just be the technique that brings FCS into the mainstream of microscopy.

REFERENCES

Becker, W. 2010. *Advanced Time-Correlated Single Photon Counting Techniques* (Springer Series in Chemical Physics). Berlin: Springer.

Brasselet, S., Peterman, E.J., Miyawaki, A., and Moerner, W.E. 2000. Single-molecule fluorescence resonant energy transfer in calcium concentration dependent Cameleon. *Journal of Physical Chemistry B*, 104, 3676–3682.

Carlsson, K., and Liljeborg, A. 1998. Confocal fluorescence microscopy using intensity-modulated multiple-wavelength scanning (IMS): Evaluation of results from spectral and lifetime imaging. *Proceedings of SPIE*, 3261, 30.

Chang, C.-W., Sud, D., and Mycek, M.A. 2007. Fluorescence lifetime imaging microscopy. In *Digital Microscopy* (Methods in Cell Biology, vol. 81), G. Sluder and D.E. Wolf, eds. San Diego, CA: Elsevier, pp. 495–524.

Cox, G.C., Matz, M., and Salih, A. 2007. Fluorescence lifetime imaging of coral fluorescent proteins. *Microscopy Research & Technique*, 70, 243–251.

Digman, M., and Gratton, E., 2007. The RICS method: Measurement of fast dynamics in cells with the laser scanning microscope. In *Image Analysis in Medicinal Microscopy and Pathology*, H.-S. Wu and A. Einstein, eds. India: Research Signpost, pp. 1–17.

Elangovan, M., Horst, W., Chen, Y., Day, R.N., Barroso, M., and Periasamy, A. 2003. Characterization of one- and two-photon excitation energy transfer microscopy. *Methods*, 29, 58–73.

Förster, Th. 1948. Zwischenmolekulare Energiewanderung und Fluoreszenz. *Annalen der Physik*, 437, 55.

Gerritsen, H.C., Asselbergs, M.A.H., Agronskaia, A.V., and Van Sark, W.G. 2002. Fluorescence lifetime imaging in scanning microscopes: Acquisition speed, photon economy and lifetime resolution. *Journal of Microscopy*, 206, 218–224.

Hess, S.T., Huang, S., Heikal, A.A, and Webb, W.W. 2002. Biological and chemical applications of fluorescence correlation spectroscopy: a review. *Biochemistry*, 41, 697–705.

Jovin, T.M., and Arndt-Jovin, D.J. 1989. FRET microscopy: Digital imaging of fluorescence resonance energy transfer. Application in cell biology. In *Cell Structure and Function by Microspectrofluometry*, E. Kohen, J.S. Ploem, and J.G. Hirschberg, eds. Orlando, FL: Academic Press, pp. 99–117.

Magde, D., Elson, E.L., and Webb, W.W. 1972. Thermodynamic fluctuations in a reacting system: Measurement by fluorescence correlation spectroscopy. *Physical Review Letters*, 29, 705–708.

FURTHER READING

Periasamy, A., and Clegg, R.M., eds. 2009. *FLIM Microscopy in Biology and Medicine*. Boca Raton, FL: CRC Press.

Periasamy, A., and Day, R., eds. 2005. *Molecular Imaging: FRET Microscopy and Spectroscopy* (Methods in Physiology Series). Oxford: Oxford University Press.

16 Evanescent Wave Microscopy

THE NEAR-FIELD AND EVANESCENT WAVES

TOTAL INTERNAL REFLECTION MICROSCOPY

Total internal reflection fluorescence microscopy (TIRF) makes use of the quantum nature of light to form a fluorescence image under conditions that are forbidden by conventional ray optics (Axelrod, 2001). The image comes from a very thin layer at the top surface of a sample and thereby offers resolution in the axial (Z) direction which confocal microscopy cannot match. To understand how TIRF works, we first must understand total internal reflection, so here is a quick refresher. We have already encountered Snell's law: sin i/sin r = n, where n is the refractive index, and i and r are the angles made to the normal by the incident and refracted rays. In this formulation, Snell's law applies going from a medium of refractive index 1 (i.e., air or vacuum) into a medium of refractive index n. Because light ray paths are generally reversible, Snell's law can be applied in reverse to deal with a ray passing from a denser medium (high n) into a less dense one. Of course, the less dense medium need not be air. It could be almost anything but for our purposes the most likely case is water (n = 1.3, or a bit more if anything is dissolved in it). So, it might be worth generalizing the formula as follows:

$$\sin\theta^1/\sin\theta^2 = n^2/n^1$$

where the suffixes 1 and 2 refer to the lower and higher refractive index media, respectively. Thus, going from air into glass n^1 is 1, and the simple form of the equation applies. If we are going the other way, from glass into air, the exit angle will always be larger than the entry one, and eventually it will reach 90°, where the emerging ray travels parallel to the glass surface. Any higher value for θ^2 would make θ^1 larger than 90°, so the ray cannot exit from the glass. Instead, it is reflected, following the normal laws of reflection, and this is called total internal reflection. At what angle will this happen? If $\theta^1 = 90°$, then sin $\theta^1 = 1$, so this is pretty simple to work out. This value of θ^2 is called the critical angle, θc, and is given by

$$\sin \theta c = n^1/n^2$$

Going from glass into air, this works out to $\sin^{-1}$ (1/1.5), which is 42°. This calculation explains why we can use 90° prisms as perfect mirrors (for example, in a binocular head). In this case $\theta^2 = 45°$, which is greater than θc, so all the light is reflected.

Although the reflection is normally total (100%), because of the inevitable quantum uncertainties, some light will exist as a so-called evanescent wave, which decays exponentially from the interface. In practical terms, it will be significant up to a distance of ~100 nm and anything present in this zone can interact with the evanescent wave. If this "anything" is a fluorochrome molecule, then it will fluoresce, but another molecule 200 nm from the surface will not. This level of depth discrimination much exceeds the optical sectioning capability of a confocal microscope.

As cell biologists, we are more likely to be interested in things in water than things in air. So how do the numbers work out for light going from glass into water? Again, it is simple:

$$\sin \theta c = n^1/n^2 = 1.3/1.5$$

so that the critical angle θc is 60°. Physiological liquids, with a fair quantity of solutes, have a higher refractive index, and if we take that index as 1.35, then sin θc will be 0.9, making θc 64°. The conclusion, therefore, is that if we illuminate with light at 64° from the vertical (26° from the glass surface), we excite fluorescence only in a very thin layer, about 100 nm deep. We will see anything in this region with very high contrast and with no contribution from out-of-focus light. One might expect, from the idea of an evanescent wave, that this image would be dim, but not so. The proportion of the light in the evanescent wave is constant; as soon as some is removed (by exciting fluorescence) more becomes present, so excitation is very efficient. The image is actually brighter than a conventional fluorescence image since the excitation is confined to a very thin layer.

How can we do this in practice? There are two approaches, illustrated in Figure 16.1. One (16.1A) is to mount the specimen on a prism with sides that make an angle steeper than θc, so that light coming through the sides is totally internally reflected. We then image from above using a water-immersion objective. Lab systems that work in this way have been built, but the requirement of putting the samples directly on the prism is inconvenient and limiting. Also, we are imaging through the whole thickness of the sample, which can lose us some light.

It is much more convenient if we can put our sample on a conventional coverslip, and commercial systems therefore use the layout shown in Figure 16.1B. Now we have epi-illumination, but instead of using the whole of the objective pupil to illuminate the sample, we bring in the light in at the extreme edge of the back focal plane. This corresponds to the maximum angle the lens can accept or produce (Chapter 1); if that angle exceeds θc, we will have total internal reflection where the coverslip meets the medium. This implies a lens of very high numerical aperture (NA). Sin θ must be greater than 0.9, and we use oil immersion to keep the angle the same all the way to the coverslip, so the NA must exceed 1.35 (0.9 × 1.5). Manufacturers make special oil-immersion lenses for TIRF with an NA of around 1.45, so that we can get a decent amount of light into the high-angle region. Olympus has gone even further and made a "super-TIRF" lens designed for a refractive index of 1.7, which can achieve an NA of 1.65. This refractive index implies using special high-index coverslips (which are expensive and can be hard to get) and methylene iodide as the immersion medium.

A

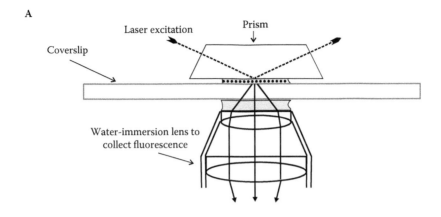

Laser excitation   Prism

Coverslip

Water-immersion lens to
collect fluorescence

B

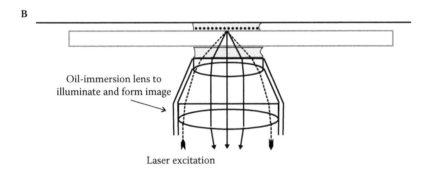

Oil-immersion lens to
illuminate and form image

Laser excitation

FIGURE 16.1 TIRF arrangements. (A) Prism system, with the sample on the face of the prism and the laser illuminating it from the back. The objective lens just collects the image. (B) Illumination through the extreme edge of a very high NA objective, so the TIRF takes place at the surface of the coverslip and the objective also collects the image. If we shift the laser beam closer to the optical axis, TIRF will not take place and a conventional epifluorescence image will be formed.

To get a lot of light into a small region requires a laser, and the laser light is coupled in just behind the objective. The position is adjustable, and in practical operation one moves the laser spot outward across the back focal plane (BFP) until TIRF is achieved. The change in the image at the point is quite striking, as the background and all out-of-focus haze suddenly disappear, and we see just the structures illuminated by the evanescent wave (Figure 16.2).

TIRF microscopy is limited to looking at the very edge of cells—the membrane and immediately adjacent cytoplasm—but this is a region of great interest to most cell biologists. The technology is neither difficult nor complicated (although the objectives are expensive) and it is rapidly becoming a routine tool in the cell biology lab. And it is the only case in which looking at a sample in water with an oil-immersion lens is permissible!

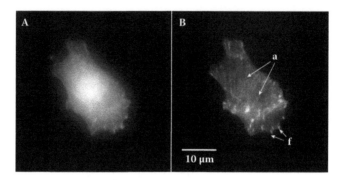

FIGURE 16.2 MTLn3 breast cancer cells stained with FITC-phalloidin to label actin under TIRF microscopy. (A) Laser set so that θ < θc; out-of-focus blur effectively hides all useful information. (B) In TIRF mode, θ > θc. Focal adhesions, f, are much more clearly demarcated at the leading edge of the cancer cell, and actin stress fibers, a, are now visible in the cytoplasm close to the coverslip. (Courtesy of Lilian Soon.)

NEAR-FIELD MICROSCOPY

What happens if we move our specimen over a tiny pinhole, much smaller than the wavelength of light? The idea was first proposed by Synge in 1928, but he had no way to implement it. Theory tells us that within the hemisphere defined by the diameter of the hole, we will be in the near-field region where light behaves as an evanescent wave that does not propagate. So if our hole is both very small and very close to the specimen, the idea should work.

The coming of the scanning probe microscope around 1980 provided the technology to make it happen. Figure 16.3 shows the basic idea reduced to its simplest form. A pyramidal tip is scanned over a sample and gradually brought into closer contact.

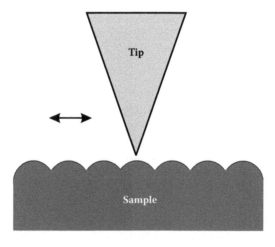

FIGURE 16.3 The basic principle of a scanning probe microscope. A pointed tip, controlled to great precision, scans over the surface of a sample.

Eventually, the interaction will be governed by forces acting at the atomic level. On this scale there will always be one atom on the tip that stands above its neighbors and therefore will be the first to interact. The key to making a useful microscope is to detect this interaction and use it to control the scanning tip.

There are various possible interactions. The original version, the scanning tunneling microscope, used electrical conduction. If the tip carries a small voltage and the sample is a conductor, then when the tip is close (<1 nm) to the surface, the quantum uncertainty in the location of electrons means that some will jump the gap and a tiny current will pass. By sensing and measuring this current, we can measure roughness on the surface down to the scale of single atoms. This highest-resolution form of SPM is of little use to the biologist because our samples are typically not conductive. The atomic force microscope (AFM) measures instead the interactions between the atoms of the sample and those of the tip. In general terms, as the tip approaches a sample, it first experiences an attractive force due to van der Waals forces. As it comes closer, it experiences a force away from the sample as tip and sample come into contact. The scanning tip is mounted on a flexible cantilever, and the force is measured from the deflection of the cantilever by using a beam of laser light reflected from the end above the tip. The AFM is capable of atomic resolution (Figure 16.4), although it is often used at larger scales.

The scanning probe gives us a platform on which to construct a near-field microscope. Figure 16.5 shows a typical design. The microscope probe is now an optical fiber, drawn out to a very fine (<100 nm) point at the tip and coated with aluminium so that light can escape only at the tip. Laser light passes along the fiber, and the tiny hemisphere at the tip defines the near-field region, which can excite fluorescence from any molecule within it. That fluorescence is collected by an oil-immersion objective lens beneath the coverslip that carries the specimen. The objective can also be used to form a conventional widefield fluorescence image of the sample. This is variously called a near-field scanning optical microscope (NSOM) or scanning near-field optical microscope (SNOM).

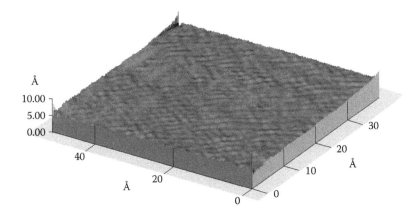

FIGURE 16.4 The surface of freshly cleaved mica, imaged at atomic resolution in the atomic force microscope (contact mode).

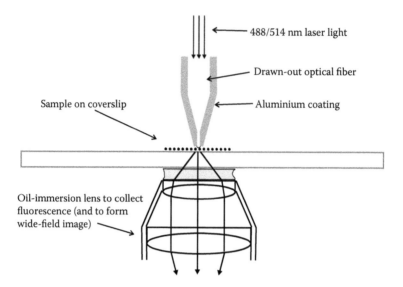

488/514 nm laser light

Drawn-out optical fiber

Sample on coverslip

Aluminium coating

Oil-immersion lens to collect
fluorescence (and to form
wide-field image)

FIGURE 16.5 Typical NSOM layout. A drawn-out optical fiber tip scans the sample and an oil-immersion lens collects the fluorescence.

In principle, we could use the system in reverse and illuminate the specimen with the objective, picking up fluorescence with the tip. This is not such a good layout, because we will be illuminating the sample and bleaching it, while we are not collecting data. Further, the fiber tip will not pass much light, so it is better for it to carry the powerful laser beam than to detect the relatively dim fluorescence.

The tricky part is controlling the tip position. Because our fluorescence is not continuous, we must use some other signal, so an NSOM is also an AFM. Contact-mode AFM (Figure 16.4) is the highest-resolution option, but the sideways forces exerted by the scanning tip often displace delicate biological specimens. This leaves us with two alternatives:

1. The attractive regime. In this method, the tip remains >1 nm from the sample (still within the evanescent wave region). In this mode, the tip is vibrated and the force is detected by its effect on the vibration frequency. The attractive regime can work quite well with the tip in air, though the AFM image will be at lower resolution than in contact mode. In water, however, the force is too weak relative to the drag of the water and control is unreliable.
2. Tapping, or intermittent contact mode. Between each step, the tip is retracted and moved, and then brought in again until it is sensed to be in contact mode. Only then is the optical image of that point collected. Because no lateral force is exerted on the sample, it generally is not affected by the contact. This mode is relatively slow.

Figure 16.6 shows the resolution that can be achieved on a dry sample: DNA labeled with in situ hybridization. Full width at half maximum (FWHM) figures of around 80 nm are obtained in conjunction with nanometer-level resolution in

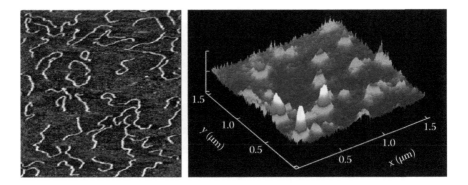

FIGURE 16.6 DNA labeled by fluorescent in situ hybridization (FISH) with single rhodamine molecules. The molecules are imaged with ~80-nm resolution. Shear-force (vibrating tip) control; the shear-force image of the DNA is shown on the left. (From van Hulst, N.F., Garcia-Parajo, M.F., Moers, M.H.P., Veerman, J.-A., and Ruiter, A.G.T., 1997, *Journal of Structural Biology*, 119, 222–231. Courtesy of Niek van Hulst.)

the AFM image. Figure 16.7 shows cell-surface lectins on a dendritic cell in water. A clever technique allows recognition of single molecules of the Cy3 label. Two channels in opposite polarizations are collected. Because individual molecules are dipoles, the fluorescence from a single molecule will be polarized and hence will appear in one or other channel, either as red or green. When several molecules are imaged together, all dipole orientations are present and there is no overall polarization so both channels will be the same signal. Many of the dots are seen to be single molecules, and a resolution of around 90 nm is obtained.

All the major SPM (scanned probe microscope) makers now offer NSOM modules based on this technology. However, its penetration into mainstream cell biology has been limited. Partly, this is a consequence of the restriction on suitable samples, but there is also a more fundamental limitation. As the size of the scanning aperture is reduced, there is an exponential decrease in the intensity of the emitted evanescent wave. This means that signal-to-noise considerations make resolution much below 100 nm difficult to achieve. Far-field super-resolution techniques (Chapter 17) can easily meet or exceed this value with fewer limitations on specimens.

An alternative approach that bypasses this restriction is now the subject of active research: apertureless NSOM (ANSOM) or tip-enhanced near-field optical microscopy (TENOM). (As so often, both acronyms refer to the same technique.) The basis of this is the well-known principle that electric charge becomes concentrated at a sharp point. Light is electromagnetic radiation, and a focused beam of light has an associated electric field, which (if the polarization direction is correct) will produce a charge on a small tip, a charge that will concentrate at the sharpest point. This, in turn, will focus the light so that it, too, is strongest at the tip. Since we are talking about a tip with a radius a couple of orders of magnitude smaller than the wavelength of light, we are again in the near-field/evanescent wave region.

In practical use, the tip is maintained close to the sample by one of the methods discussed earlier, while the sample itself is scanned by a piezo stage. Illumination and detection are by a conventional high-power objective (Figure 16.8). The complication

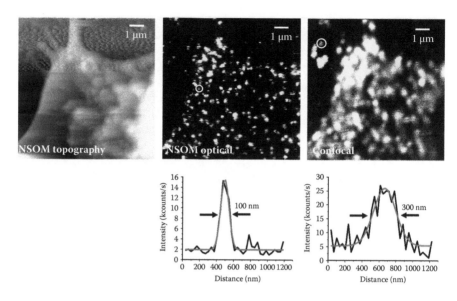

FIGURE 16.7 NSOM images of dendritic cells, in water, labeled with a Cy3 tagged antibody against the cell surface lectin DC-SIGN. Left: Topographic image. Center: NSOM image taken in two channels at opposing polarization. Red or green dots are in one channel only and are therefore single molecules. Profile of circled spot below. Right: Confocal image of the same area and profile of circled spot. (From Koopman, M., Cambi, A., de Bakker, B.I., Joosten, B., Figdor, C.G., van Hulsta, N.F., and Garcia-Parajo, M.F., 2004, *FEBS Letters,* 573, 6–104. Courtesy of Niek van Hulst.)

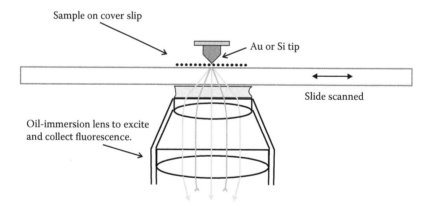

FIGURE 16.8 Diagram of a TENOM (or ANSOM). The objective lens focuses light on the tip, which is controlled in height by standard AFM methods. At the very point an intense evanescent wave is formed that excites fluorescence, collected by the same objective. To form an image the slide is scanned. If the tip is withdrawn a standard scanned image will be formed.

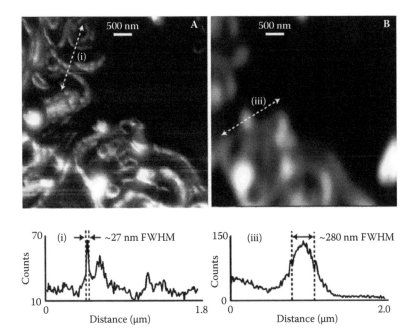

FIGURE 16.9 J-aggregates of pseudoisocyanine dye imaged under two-photon excitation at 833 nm. (A) ANSOM image using a shaped gold tip. The trace below shows a resolution of ~27 nm. (B) Far-field image with the tip withdrawn showing a resolution of ~280 nm. Imaging aggregates rather than single molecules minimises quenching by the gold tip. (Reproduced by permission from Nowak, D.B., Lawrence, A.J., and Sánchez, E.J. 2010. Apertureless near-field/far-field CW two-photon microscope for biological and material imaging and spectroscopic applications. *Applied Optics*, 49, 6766–6771.)

is that the near-field fluorescence image generated by the tip will be superimposed on a conventional scanned fluorescence image, and somehow the two must be separated. There are various approaches to this. One is to use two-photon excitation. Since the fluorescence then depends on the square of the excitation energy, this automatically enhances the near-field contribution (Nowak et al., 2010; Sánchez et al., 1999). Another is to use tapping mode and collect the signal via a lock-in amplifier that will specifically pick up the signal when the tip is in range (Mangum et al., 2008). Either way, we end up with both a conventional scanned image and a TENOM image (Figure 16.9).

There are other complications. The field enhancement is greatest if we use a gold tip, but a gold tip close (~2 nm) to the sample will tend to quench fluorescence (Sánchez et al., 1999). Maintaining the tip around 5 nm from the sample would avoid this, but this is a dead zone for AFM, between the attractive and repulsive force regimes. More conventional silicon tips do not have this problem but offer less field enhancement (Mangum et al., 2008).

Serious biological applications of TENOM/ANSOM have yet to be demonstrated, but images of dye aggregates show a resolution of 27 nm, a 10-fold increase on the far-field resolution (Figure 16.9). This far exceeds anything achieved with aperture-based NSOM, and further technical developments could well make this resolution usable in biology. Only apertureless methods seem to offer a chance for near-field microscopy to match or exceed the performance of the latest far-field techniques (Chapter 17).

REFERENCES

Axelrod, D. 2001. Total internal reflection fluorescence microscopy in cell biology. *Traffic*, 2, 764–774.

Koopman, M., Cambi, A., de Bakker, B.I., Joosten, B., Figdor, C.G., van Hulsta, N.F., and Garcia-Parajo, M.F. 2004. Near-field scanning optical microscopy in liquid for high resolution single molecule detection on dendritic cells. *FEBS Letters,* 573, 6–10.

Mangum, B.D., Mu, C. and Gerton, J.M. 2008. Resolving single fluorophores within dense ensembles: Contrast limits of tip-enhanced fluorescence microscopy. *Optics Express*, 16, 6183–6193.

Nowak, D.B., Lawrence, A.J., and Sánchez, E.J. 2010. Apertureless near-field/far-field CW two-photon microscope for biological and material imaging and spectroscopic applications. *Applied Optics*, 49, 6766–6771.

Sánchez, E.J., Novotny, L., and Xie, X.S. 1999. Near-field fluorescence microscopy based on two-photon excitation with metal tips. *Physical Review Letters* 82, 4014–4017.

Synge, E.H. 1928. A suggested method for extending the microscopic resolution into the ultramicroscopic region. *Philosophical Magazine*, 6, 356.

van Hulst, N.F., Garcia-Parajo, M.F., Moers, M.H.P., Veerman, J.-A., and Ruiter, A.G.T. 1997. Near-field fluorescence imaging of genetic material: toward the molecular limit. *Journal of Structural Biology*, 119, 222–231.

17 Beyond the Diffraction Limit

By the end of the 19th century, optical microscopy had hit the wall. Objectives had been made that were sufficiently good to achieve the theoretical maximum resolution defined by the wavelength of light. There seemed to be no obvious way forward. One approach was to reduce the wavelength, and ultraviolet microscopes had a certain vogue in the first half of the 20th century. The problems with these were many; recording the image on film is easy, but focusing on a fluorescent screen is not so easy. Lenses and coverslips had to be made of quartz, because most glass does not transmit far enough into the ultraviolet (UV) to be useful. Without the range of special glasses available for the visible range, lens correction was not so good. And cells were always going to be dead—at least pretty soon! So UV microscopy was essentially a technique for fixed material.

Electron microscopes, first developed in the 1930s, made the UV microscope obsolete. If you are going to be restricted to fixed material, why not go for a hugely shorter wavelength and two or three orders of magnitude resolution improvement? Much of what we know about the makeup of the cell has come from electron microscopes in the 50 years since biologists worked out how to prepare specimens for the new instrument. Now, electron microscopes are taken for granted. But they do not get us anywhere with living cells, so the idea of doing better with visible light has remained alive.

The coming of confocal microscopy (Chapter 5) brought a modest resolution improvement to visible light microscopy. This development also acted as a catalyst; suddenly the wall did not seem so solid after all. Since that breakthrough, several approaches to beating the Rayleigh limit have been developed, most of them depending on scanning and fluorescence. We have already discussed techniques that operate in the near field (Chapter 16) and have seen that while in principle they are not subject to the diffraction limit, in practice they have many limitations. If we can operate in the "far field," the regime of a normal microscope, we will have a much more versatile instrument. This chapter we will cover such techniques, focusing particularly on those that are available on the market to cell biologists.

4PI AND MULTIPLE-OBJECTIVE MICROSCOPY

Is there any possibility of increasing the numerical aperture (NA) above 1.4 or so? If there were, we would get better resolution without violating Rayleigh's or Abbe's criteria. One obvious possibility is to use media with a refractive index higher than 1.515, and the best candidate for this is methylene iodide ($n = 1.7$). This approach had some vogue in the early 20th century, and now that Olympus is making such lenses

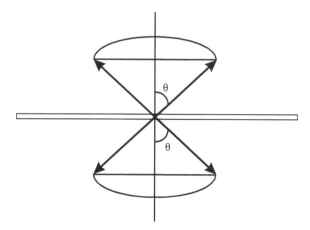

FIGURE 17.1 Two objectives on opposite sides of the specimen, the principle of the 4Pi microscope.

and matching coverslips for total internal reflection microscopy (TIRF), the possibility exists to get up to an NA of 1.65 by mounting a fixed specimen in methylene iodide. The gain in resolution, however, is only a modest 10%.

Several researchers have explored the approach of using more than one objective to obtain the equivalent of a higher NA (Figure 17.1). The most successful implementation of this idea, to date is the 4Pi microscope proposed by Cremer and Cremer (1971) and developed by Stefan Hell and colleagues (Dyba and Hell, 2002) and which is now marketed by Leica. This microscope has two immersion objectives: one above and one below the specimen. For this configuration to increase resolution, the wavefronts from both objectives must be exactly in phase with each other, and achieving this is quite a tricky task.

In principle, one can use either one-objective excitation and two-objective detection or vice versa but to get the maximum resolution benefit one should both excite and detect in phase with both. Two-photon excitation simplifies the situation, since the resolution is defined only by the excitation; it also avoids the complexity of aligning two pinholes. This approach is shown in Figure 17.2, a diagram of the Leica commercial system. The result is a microscope with much improved resolution, particularly in the Z direction, and the image is normally scanned in XZ mode. The point spread function (PSF), however, is peculiar, with large side lobes (Figure 17.3). Hence we can expect resolution of the order of 100 nm if we can get rid of the side lobes. Fortunately, the relationship between the side lobes and the peak is well defined, so that deconvolution to remove them is relatively straightforward.

Setting up a sample in the microscope is a complex procedure. The lenses are carefully matched (currently only glycerol- or water-immersion lenses are supplied as matched pairs), and the sample must be mounted between matched coverslips. There follows a 68-step alignment procedure before the final image is obtained. Both lenses must be aligned with each other, both correction collars must be adjusted to match the medium and coverslips, and finally the bottom mirror (Figure 17.1) is adjusted to get the phase correct by looking at the PSF and adjusting to get the side lobes equal.

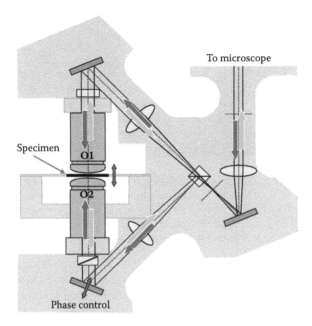

FIGURE 17.2 Diagram of the Leica 4Pi system. The sample is placed between two cover-slips, between the two objectives O1 and O2. To bring both beam paths into exact phase, the mirror beneath O2 is movable. The system fits on to a conventional Leica confocal microscope. (Courtesy of Leica Microsystems.)

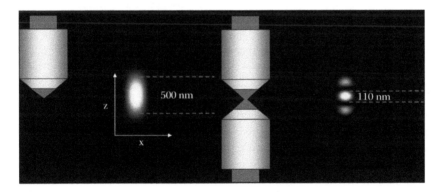

FIGURE 17.3 XZ view of the PSF of a 4Pi microscope (right) compared to that of a confocal microscope (left). (Courtesy of Leica Microsystems.)

(The inclusion of fluorescent beads with every sample is mandatory.) The results obtainable on a real sample—microtubules in a HeLa cell—are shown in Figure 17.4. The raw image has a strange appearance, thanks to the ghost images from the side lobes, but after deconvolution the gain in resolution is apparent and impressive.

Various other configurations of multiple objectives are possible. Ernst Stelzer and Steffan Lindek in Heidelberg have explored many of these. One simple idea is to

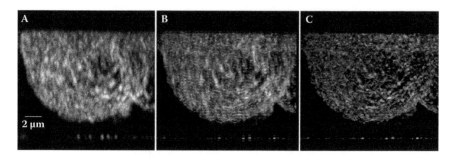

FIGURE 17.4 HeLa cell. Immunostaining of α-tubulin (Alexa488). Maximum projections from series of XZ scans. (A) Regular confocal, (B) 4Pi without deconvolution. Note the multiple image effect from the side lobes. (C) 4Pi after deconvolution, giving much better resolution than confocal. The line across the bottom is a layer of fluorescent beads to show the PSF. (Micrographs by Tanjef Szellas and Brian Bennett. Courtesy of Leica Microsystems.)

have the two objectives at right angles rather than facing each other, with the aim of achieving a PSF that is similar in all directions. From there one can add more objectives; Stelzer's group has gone as far as four in a tetrahedral arrangement. The most extreme must be a configuration with six objectives proposed by Olivier Haeberlé in France. Imagining the sample as a cube, there is one objective on each face of the cube. The advantage of using four or six lenses is that with objectives of relatively modest NA, one can achieve very high resolution with a fairly large working volume for the specimen. However, alignment would be a nightmare, and most of these proposed schemes have not been built.

With 4Pi imaging, commercial far-field microscopes have achieved resolutions comparable to those obtained with commercial near-field imaging, without its limitations. Arguably, this is not beating the diffraction limit, just being creative in getting the best from it. Can we go further in the far field?

STIMULATED EMISSION DEPLETION (STED)

Currently, the ultimate state of the art in true optical resolution lies with a clever approach to PSF engineering proposed independently by Hell and Wichmann (1994) and Baer (1994), and developed into a practical tool by Stefan Hell's group: stimulated emission depletion (STED). To understand STED, we need to realize that if a molecule is excited and ready to fluoresce, we can force that fluorescence to occur immediately by hitting the molecule with light at the emission wavelength. That is how a laser works: We have a population of atoms or molecules in an excited state, and the photons moving back and forth in the laser cavity force them to decay to the ground state and emit photons in their own right (Chapter 5). The same principle applies to any excited molecule: we can force it to fluoresce by hitting it with a photon at its emission wavelength. The emitted photon will always be at the same wavelength as the photon causing the depletion (Figure 17.5).

When our scanning spot in confocal or two-photon microscopy hits a point on the sample it will excite fluorescence in the whole volume of the PSF. Imagine now that

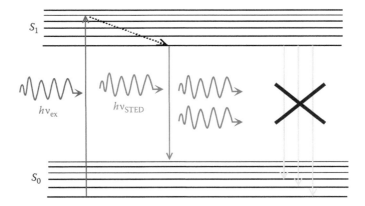

FIGURE 17.5 Jablonski diagram for STED. The arrival of a photon within the possible emission range forces immediate de-excitation with emission of another photon in phase and at the same wavelength (red). Normal fluorescence (green) is therefore inhibited. (Adapted from a diagram from Leica Microsystems.)

we also hit that spot with a ring of light at the depletion wavelength. All fluorescence in that ring will immediately be triggered and emitted at the same wavelength as the depletion beam. The remaining central part of the PSF will be unaffected and will emit over the normal fluorescence lifetime at the normal range of wavelengths. If we collect only the regular fluorescence and not the stimulated fluorescence, we will have a much smaller effective PSF and hence improved resolution (Figure 17.6).

So how do we generate this hollow PSF? There are several approaches that have been used and even more that have been proposed but never implemented. The most common is to use a spiral phase plate, that is, a piece of glass with a spiral ramp ongoing from zero to one wavelength path difference around the circle. This causes destructive interference at the center of the Airy disk so the center of the spot is dark, and all the energy is in the outer part. This will give us super-resolution in the XY plane, but in Z our resolution is still standard confocal (Figure 17.7). This is the configuration used in current commercial STED implementations.

In some of Hell's original work he used a PSF that was hollow in X, Y, and Z, but its intensity was not even in all directions, so the resolution was not isotropic. The idealized depletion PSF of Figure 17.6 is still elusive. However, by splitting the depletion beam into two parts, sending one through a spiral phase plate to give depletion in the XY plane and sending the other part through a different mask that will give illumination above and below the focus but not on it, we can come quite close (Harke et al., 2008). This currently looks like the most promising approach to taking commercial STED from 2D to 3D, though at the time of this writing no commercial implementation has been announced.

There is no actual limit to the resolution improvement. If we increase the power of the depletion laser pulse, the depletion will be saturated in the volume we already covered—there are no more electrons to de-excite—but we will push the boundary of the depletion zone further into the hole (Figure 17.8). Our PSF can become arbitrarily small; we are limited only by our ability to detect it.

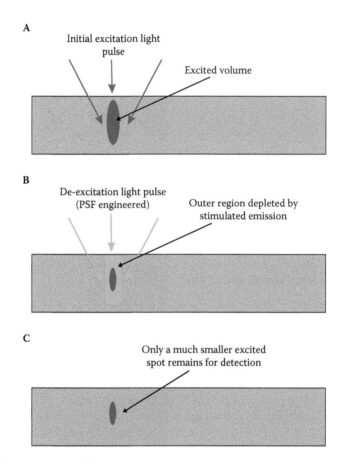

FIGURE 17.6 By clever PSF engineering, we can generate a hollow PSF that will de-excite the outer part of our excited volume.

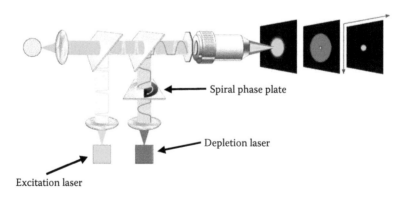

FIGURE 17.7 STED configuration using a spiral phase plate. (Courtesy of Leica Microsystems.)

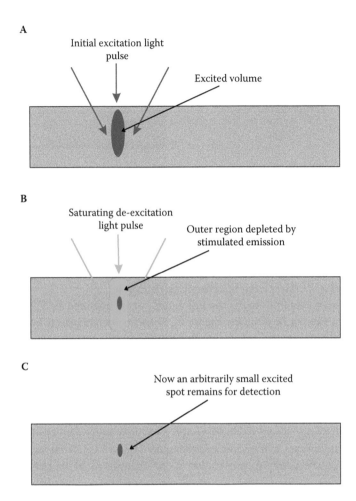

FIGURE 17.8 If we increase the depletion pulse to saturating levels, we can make our final PSF arbitrarily small.

Clearly we need to make sure we only detect the undepleted fluorescence. There are two ways we can do this. If we use pulsed lasers, we can send the depletion pulse immediately after the excitation pulse. By then detecting only after the depletion pulse, we just get the undepleted fluorescence (Figure 17.9), since depletion is instantaneous. This is the most efficient method; we can deplete at any wavelength we like to get maximum depletion efficiency. We also ensure that we will not get any light emitted from the original spot before it is depleted. The downside is the cost and complexity of synchronizing two pulsed lasers and implementing time-gated detection.

Alternatively, we can used filter-based detection. This is possible because we can deplete at any wavelength within the possible emission spectrum of the fluorochrome. If we are well off the peak, depletion is less efficient, so we need more laser power, but that is not usually in short supply. So we can detect with a bandpass filter passing everything between the excitation wavelength and the depletion wavelength,

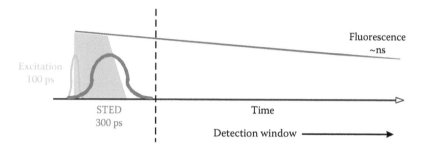

FIGURE 17.9   Time-gated detection. If we start detection after the depletion pulse we will capture almost all of the lifetime of the undepleted fluorescence (brown line), while not seeing the depleted emission (yellow). (Adapted from a diagram from Leica Microsystems.)

which will be most of the fluorescence. Since the smaller the spot, the weaker the signal, an avalanche photodiode (Chapter 4) is usually the detector of choice. At the time of this writing commercial systems are available with either pulsed or continuous wave (CW) lasers, but both use filter detection.

There is no actual limit to the resolution that can be obtained by STED, but in the end we must run out of photons. Figure 17.10 shows a comparison between confocal and STED images of FtsZ in *Bacillus subtilis* from a commercial Leica system (Jennings et al., 2010), which offers 80 nm guaranteed resolution and a spectacular improvement in image quality. Resolution of between 80 and 90 nm was measured on this sample.

In principle STED can even be done in combination with 4Pi optics. This may seem pointless since the resolution of STED is essentially unlimited, but photons are always the limiting factor, so doubling the collection angle and starting off with a smaller excitation spot will give us a further gain. Figure 17.11, from Stefan Hell's lab, shows what can be achieved: a resolution in one direction of 30 nm in far-field optics.

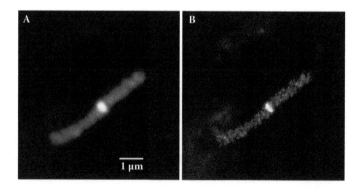

FIGURE 17.10   (A) Confocal and (B) STED images of *Bacillus subtilis* immunolabeled with Atto-647 to the cytoskeletal protein FtsZ. (Micrographs by Phoebe Jennings, Liz Harry, and the author, with thanks to Leica Microsystems for making the instrument available. See Jennings et al., 2011.)

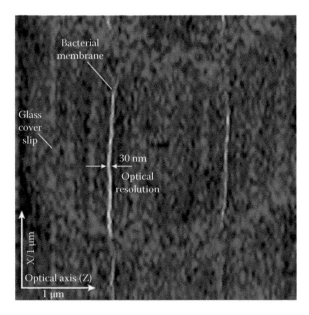

FIGURE 17.11 Membrane of *Bacillus megaterium*, imaged by 4Pi STED microscopy, showing a resolution in the Z direction of 30 nm. (From Dyba, M., and Hell, S.W., 2002, *Physical Review Letters,* 88, 163901. With permission.)

At this point, we have moved a very long way from the diffraction limit. In the real world of cell biology, the scarcity of photons available from such a small volume probably will not let us get any further. A more useful question than the ultimate attainable resolution is what we might expect to achieve with everyday tools available in our labs. In such a fast-moving field, it must be rash to make predictions, but the fundamentals of STED are relatively simple, and the components are likely to become cheaper. On the other hand, the complexity of setting up and aligning a 4Pi system will always be a daunting task. So it seems likely that single-objective STED will become a routine tool in cell biology labs.

STRUCTURED ILLUMINATION

If we illuminate a finely detailed pattern with a known pattern at similar scale we will see a Moiré pattern, which is much larger in scale (Figure 17.12). This is a well-known effect in everyday life, and its significance in microscopy is that we now have information at relatively low resolution about structure that is present at higher resolution. It is not an actual image, but since we know the illuminating pattern we can recover the original detail mathematically even if we cannot resolve it directly. This is the basis of a method of super-resolution developed principally by Mats Gustafsson (2005) at the University of California. Since any pattern we can generate is also limited by the resolution of our optical system, the pattern and our finest detail will be at a similar scale.

A   B   C

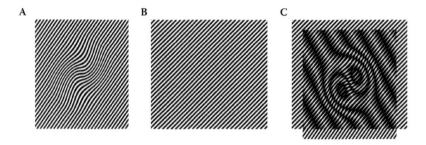

FIGURE 17.12 The principle of structured illumination. (A) A pattern with fine detail is illuminated (B) with a regular pattern at the same scale. (C) The result in a moiré pattern with detail on a much larger scale. So long as we know the pattern (B) we can reconstruct the original fine pattern (A). (From Gustafsson, M.G.L., 2005, *Proceedings of the National Academy of Science USA*, 102, 13082–13086. With permission.)

A   B

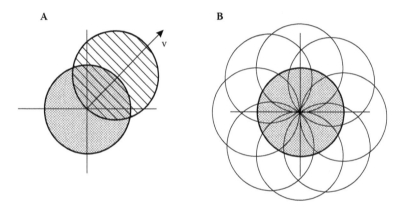

FIGURE 17.13 (A) The shaded circle shows, in reciprocal (Fourier) space, the range of detail that can be resolved by a conventional microscope, with the perimeter defining the finest level of detail. By applying a pattern drawn at the resolution limit we can shift the range of resolved detail to finer limits (striped circle), but only in one direction (v), in the process losing resolution in other directions. (B) By successively applying different orientations of the pattern we can get the same extended resolution in all directions. (Adapted from Gustafsson, M.G.L., 2005, *Proceedings of the National Academy of Science USA*, 102, 13082–13086.)

The effect of this is that if we illuminate our sample with a grid of parallel lines at the resolution limit, we will generate a pattern that can give us information about structure at half the scale of the pattern (or, in frequency terms, at twice the frequency) but only in one direction, normal to the illuminating stripes. To illuminate the whole image we will have to use two or more positions of the pattern and then we will need to repeat the process at several different angles to produce a complete image (Figure 17.13). The computer can then generate an image that will have a resolution about twice as good as a conventional image, around 100 nm (Figure 17.14).

This technique was first commercialized by Deltavision in its OMX microscope, which was used for Figure 17.14, but commercial systems have since appeared from most of the major manufacturers. Since the illumination pattern will only be in focus

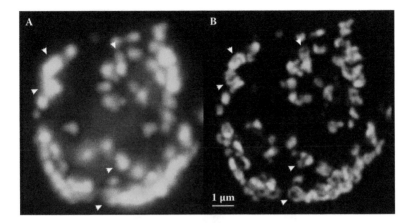

FIGURE 17.14 (A) Widefield and (B) structured illumination microscopy of red blood cells infected with the malaria parasite *Plasmodium falciparum*. The cells were labeled with an antibody recognizing structures known as Maurer's clefts. Three grating orientations were used and the image is a projection of three optical sections spaced 125 nm apart. (Modified from Hanssen et al., 2010. Courtesy of Leann Tilley and Nick Klonis.)

at one plane, super-resolution detail also only comes from one plane, so the method has inherent optical sectioning capability. Z resolution is rather better than a confocal microscope, at about 200 to 250 nm.

In principle we can increase the resolution beyond this, indeed, like STED, by a theoretically unlimited amount. This has been done experimentally but not yet commercially (Gustafsson, 2005). As with STED, the secret is saturation, in this case of the fluorescent excitation. If we increase the excitation intensity we will bring all the fluorophores in the central region of each stripe into an excited state. This will effectively "white out" this region; it will all be uniform in intensity (Figure 17.15). The dark stripes will become narrower and so will the band in which the intensity changes. Again, as with STED, we have effectively created a pattern that is finer than the resolution limit of the microscope. To make use of this we now have to take many exposures, shifting the grid at each orientation to bring the fine detail over every part of the sample—the number of images depending on the degree of saturation. Figure 17.16 demonstrates the potential of the approach using a test sample of fluorescent beads. Resolution better than 50 nm is clearly demonstrated, though to obtain this 9 exposures were taken at each of 12 orientations, a total of 108 images.

The difficulty in applying this to real biological samples is that such intense illumination tends to cause rapid bleaching of the fluorochrome. Nevertheless, it is possible with a strongly labeled specimen and a robust stain, as Figure 17.17 shows. In this case the level of saturation was lower than in Figure 17.16, and the measured resolution is only 80 nm. This still matches what commercial STED systems can reach.

STOCHASTIC TECHNIQUES

Rayleigh's criterion determines the ability to distinguish two separate objects. However, if we have just one Airy disk, we can locate its center and thus the position

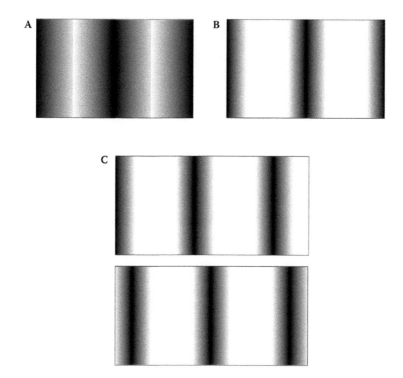

FIGURE 17.15 (A) The normal structured illumination pattern follows a sine-wave intensity profile (since it is at the limit of optical resolution). (B) If we increase the intensity to saturation we "white-out" the center of each stripe but at the same time create a pattern below the resolution limit at the edge. (C) To use this to obtain even higher resolution we must use multiple positions of the pattern with tiny shifts between each.

of the original object with an accuracy determined only by the signal to noise ratio. There is no inherent physical limit. If we can then over a period of time measure a large number of these centers, we will eventually build up a complete image point by point. If the signal from each point is strong enough, we can attain an extremely good resolution; 20 nm is often claimed. This is the basis of a range of essentially similar techniques known by a bewildering number of acronyms, including PALM (photoactivation localization microscopy; Hess et al., 2006; Gould et al., 2009), STORM (stochastic optical reconstruction microscopy), d-STORM (direct STORM), and GSDim (ground state depletion imaging).

The key is to be able to image isolated Airy disks, each deriving from an individual fluorochrome molecule. To do this we have to have a system in which most of the fluorochrome molecules in the sample are turned off—nonfluorescent—but can be slowly returned to a fluorescent state. If this process is slow enough, the molecules will "switch on" at random times and at any moment there will only be a sparse population emitting light. We record these in a widefield image with a high-sensitivity CCD camera (Chapter 4), then record the scene again a few moments

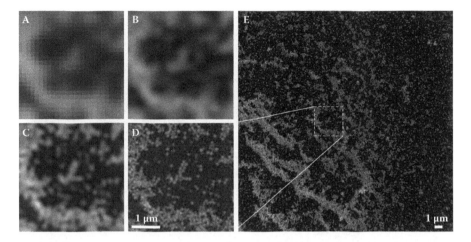

FIGURE 17.16 Demonstration of saturated pattern microscopy using 50 nm fluorescent beads. (A) Conventional image acquired with Nyquist sampling. (B) The same image "de-pixelated." (C) The same area imaged with conventional structured illumination. (D) The same area imaged with saturated structured illumination. (E) The full image area under saturated pattern illumination. (From Gustafsson, M.G.L., 2005, *Proceedings of the National Academy of Science USA*, 102, 13082–13086. With permission.)

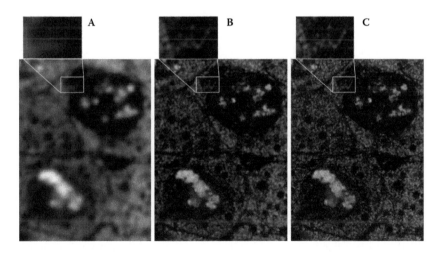

FIGURE 17.17 Saturated pattern microscopy of *Drosophila* embryo stained with Pro-Pro-3 for DNA and RNA. (A) Conventional fluorescence image, estimated full width at half maximum (FWHM) of 280 nm. (B) Linear structured illumination, average of four data sets, estimated FWHM 110 nm. (C) Saturated structured illumination, average of four data sets, estimated FWHM 80 nm. (Courtesy of Mats Gustafsson.)

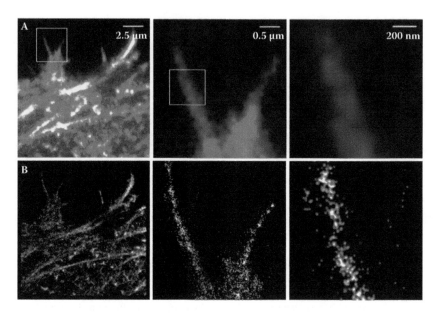

FIGURE 17.18 HeLa cells labeled with TD-Eos conjugated to LifeAct (actin probe). (A) Sum of all the acquired images, equivalent to a conventional fluorescence image, at three scales. (B) Plot of all the calculated centroids at the same scale. (Image from Katharina Gaus, in collaboration with Will Hughes, James Burchfield, Jamie Lopez, and David James.)

later when a fresh group of molecules is fluorescing. We will capture hundreds, or even thousands, of images to record just one field of view. If we sum all these images they will (or should) give us a conventional fluorescence image, which is useful for quality control (Figure 17.18A). A computer program then analyzes each image. Any spots that are too dim are rejected as random noise, any that are too bright are rejected as being more than one Airy disk, and for the remainder a spot is drawn at the calculated center of the disk. The size of this spot should reflect the accuracy expected from the signal–noise ratio. Plotting all these spots together gives us our super-resolved image (Figure 17.18B).

A concrete example based on PALM might make this easier to follow. We have a sample expressing the fluorescent protein Eos, which in its native state fluoresces green when illuminated by blue light but is nonfluorescent in green light. After activation by short-wavelength light it fluoresces red with green excitation. So we give our sample a short burst of 405 nm light, which will activate a small proportion of the Eos molecules. We then illuminate the sample with green light (561 nm) to excite the red Eos fluorescence and collect a sequence of images until the image is dark; all the activated molecules have been bleached. (Since we are imaging single molecules, they do not fade, they switch off.) Then we repeat the burst of 405 nm light and excite a fresh batch of Eos molecules. We continue doing this until there are no more Eos molecules left to activate. This is essentially how Figure 17.18 was created, except that in this case both activation and excitation were continuous. The field of view was continuously irradiated with 405 nm light at 1% of the laser power to give suitably slow photoactivation. At the same time it was illuminated with 100% of the 561 nm laser to give bright fluorescence followed by rapid bleaching.

STORM, in its original version, depended on labeling with two linked fluorochromes such as CY3 and CY5 (Bates et al., 2007). The shorter wavelength dye served to activate the longer wavelength one, which produced the image. Longer wavelength light pushed the complex back into a dark state, from which point the cycle could be repeated. d-STORM and GSDim simplify the process by using a single dye (and many routine fluorochromes are suitable) that can be driven into a dark state (Heilemann et al., 2009). This is variously described as a "long-lived triplet state" and a "radical anion," but it seems likely that in each implementation both are in fact involved, the precise mix depending on the chemical environment. The chemical makeup of the medium is very important, with oxygen levels closely controlled to maintain the fine balance between permanent bleaching and reactivation. In some versions reversion to the fluorescent state is spontaneous, in others it is driven by short wavelength (405 nm) light. In all of these variants the process is reversible, so that each fluorochrome molecule can contribute multiple times to the final image before it becomes irrevocably bleached. This has been claimed to offer more photons per molecule than are obtained in PALM, but this is debatable since most activatable fluorescent proteins are quite stable and will last a long time before bleaching.

With careful choice of wavelengths two-color PALM is possible (Shroff et al., 2007). Both fluorescent proteins must be activated by the same wavelength but have different excitation wavelengths. Figure 17.19 illustrates this. Both TD-Eos and PS-CFP are activated at 405 nm, but CFP is excited at 488 nm and Eos at 561 nm, so with sequential excitation both can be imaged. STORM is equally possible in two colors (Bates et al., 2007).

An extension of this technique allows for three-dimensional reconstruction (Huang et al., 2008). If we have enough photons available in our reconstruction we can determine whether our Airy disk is in focus or out of focus, and in the latter case by how much. An unaberrated PSF is symmetrical about focus (Chapter 1) so there remains an ambiguity about whether we are over or under focus. There are two ways to resolve this ambiguity. One is to introduce an aberration—spherical aberration or on-axis astigmatism—so that the PSF is different above and below focus. This does

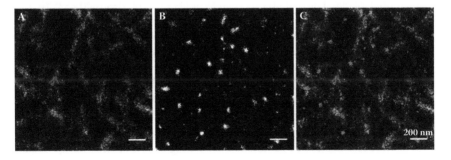

FIGURE 17.19 Dual-color PALM. (A) Life-Act (actin label) with PS-CFP. (B) Glut-4 (glucose transporter) with TD-Eos. (C) The two images merged. (Image from Katharina Gaus, in collaboration with Will Hughes, James Burchfield, Jamie Lopez, and David James.)

have the disadvantage that the in-focus spot will be enlarged, compromising the quality of the reconstruction, so some resolution may be lost. The other approach is to take images at two planes, so that the direction of defocus is known. Both images are taken at once; a beamsplitter splits the signal between two cameras, placed at slightly different focal positions. This still gives a perfect Airy disk at focus, but since each image only gets 50% of the photons, there is a trade-off in the signal-to-noise ratio. Both methods trade off some accuracy, but the ability to create a super-resolved 3D reconstruction without taking a through-focal series is something no other super-resolution scheme can offer.

So where does this place stochastic methods in terms of resolution in biological systems? The effective resolution is not calculable by conventional methods. Localization accuracy has been estimated at 5 to 20 nm for 500 to 3000 photons from a given molecule (Heilemann et al., 2009), but the distance from a fluorescently labeled antibody to its target (with primary-secondary labeling) can be as much as 15 nm, and this now becomes the resolution-limiting criterion. The other limitation is the trade-off between sparseness of labeling and time. As soon as labeling becomes dense, much of the signal has to be rejected because the PSFs of adjacent molecules overlap.

The great benefit of these stochastic techniques is that they require little specialized hardware. The camera needs to be highly sensitive, usually an EMCCD, and rigidity is at a premium, but it is much cheaper than any other super-resolution approach. Often STORM or PALM is done in TIRF mode, since this offers a very good signal-to-noise ratio, but even this is not expensive compared to the hardware for other super-resolution systems. Most microscope manufacturers offer dedicated packages, but a do-it-yourself approach is perfectly feasible, and there is public-domain reconstruction software (an ImageJ plug-in) available to do the fitting (Henriques et al., 2010).

SUPER-RESOLUTION SUMMARY

So which technique is the way to go? There is no easy answer. 4Pi is too limited in resolution and too complex to set up (changing samples takes over an hour) to be a realistic contender. STED is a strong contender. It is as simple to use as a routine confocal microscope. Current commercial systems must be regarded as early technology, and by the time this is in print 3D STED will almost certainly be a commercial product. But STED will always have limitations on usable dyes. Structured illumination has no such limitations and can offer resolution at half the Rayleigh limit quite simply. Saturated pattern, however, will probably remain just a "proof of concept" since most dyes will fade under that regime, and the sheer number of images required is quite daunting. STORM and PALM also require daunting numbers of images, but offer a very high ultimate resolution with relatively low hardware cost. Future developments will be driven by chemistry rather than physics or biology. In the end the restriction will be how dense labeling can be before pile-up (multiple molecules fluorescing within one Airy disk) becomes limiting. Images are probably destined to be forever spotty.

REFERENCES

Baer, S. 1994. Method and apparatus for improving resolution in scanned optical system. U.S. Patent #5,866,911, filed July 15, 1994, published February 29, 1996 (as International Application WO 9606369A2), issued February 2, 1999.

Bates, M., Huang, B., Dempsey, G.T., and Zhuang, X. 2007. Multicolor super-resolution imaging with photo-switchable fluorescent probes. *Science*, 317, 1749–1753.

Cremer, C., and Cremer, F. 1971. Verfahren zur Darstellung bzw. Modifikation von Objekt-Details, deren Abmessungen ausserhalb dersichtbaren Wellenlängen liegen. German patent application #P2116521, filed April 5, 1971, published October 12, 1972.

Dyba, M., and Hell, S.W. 2002. Focal spots of size $\lambda/23$ open up far-field fluorescence micros-copy at 33 nm axial resolution. *Physical Review Letters*, 88, 163901.

Gould, T.J., Verkhusha, V.V., and Hess, S.T. 2009. Imaging biological structures with fluores-cence photoactivation localization microscopy. *Nature Protocols*, 4, 291–308.

Gustafsson, M.G.L. 2005. Nonlinear structured-illumination microscopy: Wide-field fluo-rescence imaging with theoretically unlimited resolution. *Proceedings of the National Academy of Science USA*, 102, 13082–13086.

Hanssen, E., Carlton, P., Deed, S., Klonis, N., Sedat, J., DeRisi, J., and Tilley, L. 2010. Whole cell imaging reveals novel modular features of the exomembrane system of the malaria parasite, *Plasmodium falciparum. International Journal of Parasitology*, 40, 123–134.

Harke, B., Ullal, C.K., Keller, J., and Hell, S.W. 2008. Three-dimensional nanoscopy of col-loidal crystals. *Nano Letters*, 8, 1309–1313.

Heilemann, M., van de Linde, S., Mukherjee, A., and Sauer, M. 2009. Super-resolution imag-ing with small organic fluorophores. *Angewandte Chemie–International Edition*, 48, 6903–6908.

Hell, S.W., and Wichmann, J. 1994. Breaking the diffraction resolution limit by stimulated emission: Stimulated-emission-depletion fluorescence microscopy. *Optics Letters*, 19, 780–782.

Henriques, R., Lelek, M., Fornasiero, E.F., Valtorta, F., Zimmer, C., and Mhlanga, M.M. 2010. QuickPALM: 3D real-time photoactivation nanoscopy image processing in ImageJ. *Nature Methods*, 7, 339–340.

Hess, S.T., Girirajan, T.P.K., Mason, M.D. 2006. Ultra-high resolution imaging by fluores-cence photoactivation localization microscopy. *Biophysical Journal*, 91, 4258–4272.

Huang, B., Wang, W., Bates, M., and Zhuang, X. 2008. Three-dimensional super-resolution imaging by stochastic optical reconstruction microscopy. *Science*, 319, 810–813.

Jennings, P., Cox, G.C., Monaghan, L., and Harry, E.J. 2010. Super-resolution imaging of the bacterial cytokinetic protein FtsZ. *Micron*, 42, 336–341.

Shroff, H., Galbraith, C.G., Galbraith, J.A., White, H., Gillette, J., Olenych, S., Davidson, M.W., and Betzig, E. 2007. Dual-color superresolution imaging of genetically expressed probes within individual adhesion complexes. *Proceedings of the National Academy of Science USA*, 104, 20308–20313.

Appendix A: Microscope Care and Maintenance

In the cell biology lab, the major enemies of microscopes are dust, immersion oil, and aqueous solutions. As with so many things, prevention is both better and easier than a cure. There is no excuse for giving dust access to the microscope; always cover a microscope when it is not in use and always put in plugs when you remove lenses. Keep spare objectives and eyepieces in their containers. If the camera is removed, be sure to put a cap on the C-mount port.

Immersion oil is not so avoidable, but it must be handled with care. Use only as much oil as is necessary; avoid slopping it everywhere. When the slide has oil on it, be extremely careful not to swing the 40× dry lens into position; it will hit the oil and, because the front element is highly concave, that oil is very difficult to remove. Unless you are actually throwing the slides away, always clean off the oil as soon as you remove them from the microscope, even if you never intend to look at them again. You may decide later to pop a slide back in for another look. Anyway, if you do not clean up the oil, it will spread everywhere. Clean the oil-immersion lens as soon as you have finished with it, using a dry lens tissue. If you do not, before long you will deposit oil unexpectedly on a new slide and then get it on a dry lens.

On an inverted microscope, there is always the risk that oil will run down into the lens barrel. If this happens, you must send the lens back to the manufacturer for cleaning. To avoid this inconvenience, use just enough oil—no more—and remove oil-immersion lenses from the microscope after use (putting plugs into the lens sockets, of course). Store the lenses in their containers pointing downward so that any traces of oil will run out. The same applies to water-immersion lenses. (On upright microscopes, there is no need to remove the objectives, but they should still be stored pointing downward when off the microscope.) We have a rack for the objective lens boxes shaped so that they can only be put in pointing downward (Figure A.1); in any multiuser facility, it is well worth making something similar. When you put an objective back on the microscope, just light finger tightening is all that is needed. Do not tighten it with ferocious force.

The aqueous solutions used in cell biology are typically *very* corrosive to microscope parts. Remember that only dipping lenses (which usually have ceramic barrels) are intended to come into contact with physiological solutions; high numerical-aperture water-immersion lenses are intended for distilled water only. Dipping lenses must be cleaned with distilled water and then dried (using lens tissue) after use. If you do have an accident and spill liquid or break a coverslip, clean it up immediately. Soak up the liquid with a dry cloth or tissue, wipe the area with a tissue moistened (not sopping) with distilled water, and dry it carefully. Be absolutely sure you have extricated any fragments of broken glass; glass fragments that get into the stage movements or the focus rack of the condenser can do a lot of damage. If liquid has

FIGURE A.1 A simple rack to hold objective canisters so that the lenses point downward.

seeped somewhere you cannot reach, either find somebody who knows how to dismantle the part or work out how to do it yourself. Do not hope for the best; you will always be disappointed.

CLEANING

Despite all precautions, you must clean the microscope sometimes to keep it in top condition. Routine cleaning is not generally a big deal. Microscopes are basically low-tech devices; usually, you just need a couple of hexagon keys for any dismantling that the user can safely do.

Objective lenses need cleaning most often. Oil-immersion lenses usually have a fairly flat front element and therefore need little more than wiping with dry lens tissue. Dry lenses, however, are deeply concave, and if oil or saline gets on a dry lens, then removal can be difficult. The first requirement is to be able to see the problem; an eyepiece used backward is a handy magnifying glass for this purpose (Figure A.2). Be very cautious about using solvents. The first line of attack should always be a clean dry lens tissue. If it is likely that saline or other aqueous solutions could be on the lens, try distilled water next. However, the most likely culprit is always immersion oil, and a handy (if unconventional) way to clean oil from deeply recessed front elements is to break a piece of Styrofoam and press the freshly exposed (and therefore clean) surface on the lens, and then rotate it (Figure A.3). It is

FIGURE A.2   Use an eyepiece held backward as a magnifying glass to check the front face of an objective.

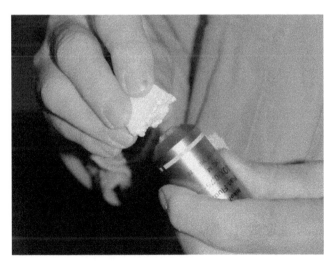

FIGURE A.3   Use a freshly fractured piece of Styrofoam to clean an objective lens.

very rare indeed that any further cleaning is needed, but if you have to use a solvent, then xylene is probably the best. Always keep the lens pointing downward so that the solvent cannot run inside, and use a lens tissue folded to a point and dipped in a little xylene. Xylene should be used in a fume hood. Dry the lens carefully with fresh lens tissue. Never use the polystyrene trick after putting xylene on a lens.

Larger optical surfaces often acquire fingerprints, and a diluted solution of window-cleaning fluid is good for removing them. This solution also works well on computer monitors. Be very cautious with mirrors: Unlike domestic mirrors, microscope mirrors usually have the coating on the front of the glass. These coatings can be very fragile; if possible, just blow off dust without touching them. Fluorescence filter cubes or sliders often acquire fingerprints because people forget that there is a filter at the back as well as the obvious one on top. The filters are safe to clean, but dichroic mirrors are not. Fortunately, in most designs the dichronic is inaccessible, but if you are changing an excitation or barrier filter take great care not to touch the dichroic. Photonic Cleaning Technologies produces an interesting cleaning product, First Contact, which is a polymer solution that is painted on to the surface and peeled off when dry. This product is claimed to be safe with front-coated mirrors and all types of lenses, but it must be kept away from plastic.

Regular dismantling and cleaning are mandatory for certain parts of a microscope. One such part is the condenser, particularly on an upright microscope where everything falls on to it. The condenser is easy to remove (Figure A.4). Typically, you move the stage coarse focus to its highest position and the condenser focus to its lowest, then undo a thumbscrew. Some condensers have swing-out top lenses; in this case the dirt can get underneath the lens as well. The stage, another item that receives a lot of abuse, is also normally easy to remove. If the stage's movements are getting stiff, then a thorough clean is in order. Be cautious with lubrication or it will get everywhere; a little petroleum jelly should suffice.

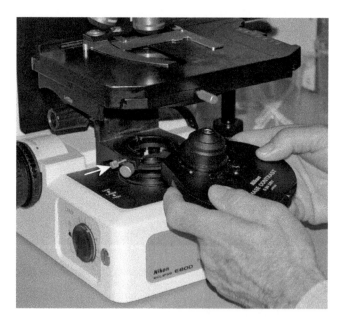

FIGURE A.4 Removing the condenser. On this Nikon microscope, the arrowed thumbscrew holds the condenser in place.

Spots of dust in the image can be very frustrating, but a little detective work can often help you discover where cleaning is needed. Do the spots stay constant with different objectives? If so, they must be in the head somewhere. If they rotate with the eyepieces the location is obvious; if not, you must look further. Removing the head is usually simple and gives you some access to optic elements (if the head is a trinocular with a slider, be sure to clean in both positions). The actual prisms, however, are pretty inaccessible, particularly the top surfaces, which is where dust accumulates if the eyepieces are left out. Blowing air down or using a clean sable brush are about the only options. If the dust is below the sample, then it must be in a plane more or less congruent with sample and image planes, which means in the illuminator rather than the condenser. If cleaning the glass where the light enters does not fix the problem, you may be able to remove this glass and clean (carefully) the mirror below. You can also easily take off the illuminator and clean the lens there. To get any further, you must lay the microscope on its side and take off the base cover, but this is usually not difficult.

THE FLUORESCENT ILLUMINATOR

High-pressure mercury lamps are temperamental things, but much of the mythology that has grown around them is inaccurate. They will not explode if used beyond their rated life; they just get dim. Neither will they explode if someone turns them on again before they have cooled down, but doing so is still a bad thing. What happens when you turn on an uncooled mercury lamp is that the arc will probably start from a fresh point, and thereafter it will flicker as it jumps between the two spots. What will make a lamp explode is operating it at the wrong voltage or frequency, so always check that these are set right on a new microscope. If a lamp does explode, evacuate the room for half an hour until everything has cooled, and expect to replace at least the mirror and maybe the lamp condenser.

The lamp is extremely strong and under no pressure when cold, but when it is hot the quartz envelope is fragile and the interior is at high pressure. So do not do anything to an illuminator unless it is cold. Also, make sure the power supply is unplugged, because the voltages used are quite high. When you change the lamp, do not touch the quartz envelope with your fingers; hold it by the metal tips or use a tissue. One connector is a rigid clamp, the other is always a flexible braid or spring, because the lamp expands a lot when operating. These connections have to take a lot of thermal cycling, so check from time to time that they are tight.

Many people seem to have difficulty centering the lamp after a change, but approached methodically, alignment is no great mystery. Remove an objective and put a piece of white paper or card on the stage. Select blue illumination, because this is the dimmest part of the mercury spectrum. If there is no mirror, just focus and center the image of the arc. A mirror doubles the useful light but gives us a bit more to align. The aim is not to focus the mirror image on to the arc itself—this may overheat it—but to have the two images side by side and centered. Figure A.5 shows the desired outcome. The confusion generally comes when we mistake the mirror image for the direct image; often only one is initially in view. Moving the lamp moves both

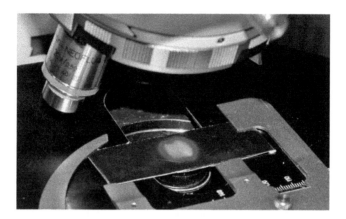

FIGURE A.5   Aligning the mercury arc. The mirror image (left) is set to be alongside the direct image (right). A black card was used to make photography simpler, but one would normally use a white card.

images but moving the mirror affects only the mirror image. Therefore, the first step is to identify the direct image by checking that the mirror alignment screws do not move it. Focus the image with the lamp focus and bring it to the correct position with the lamp alignment screws. Then focus the mirror to get a sharp mirror image and adjust the mirror alignment to bring it alongside the direct image. You may need to defocus the lamp slightly to get completely even illumination.

Appendix B: Keeping Cells Alive under the Microscope

Eleanor Kable and Guy Cox

To cover the entire topic of live cell imaging is beyond the scope of this book. In fact, the topic would be a book in itself. This appendix is intended to be an eye-opener to the possibilities and pitfalls of currently available technologies. The important thing to remember is to use the equipment you need to solve your problem. How quickly does your event take place? Is it temperature dependent? Is it time dependent? Is it pH dependent? Is it wavelength sensitive? Is it sensitive to probe concentration?

If you are working with plant, yeast, or bacterial cells, you will find that these cells are very forgiving. Fish and amphibian material is quite robust, but mammalian cells are less so. Most microscopes are housed in rooms that are held at a constant temperature of between 21°C and 24°C. This is not an optimal temperature for working with live mammalian cells. There are ways of getting around the temperature problem. Your project will determine the expense.

CHAMBERS

A vast array of environmental chambers exists. Many such chambers have been made by experimenters but more are becoming commercially available. At the simplest level, cells can be grown on coverslips, or samples placed on a slide and covered with a coverslip. A paper spacer can increase the volume and avoid crushing samples such as embryos. Sealing the slide avoids liquid loss but prevents gas exchange and may subject the sample to pressure, especially with an oil- or water-immersion lens. Leaving the slide unsealed allows evaporation and hence changes in osmolarity. If you are doing experiments over a very short time range and you use enough controls, this simple chamber is useable. Over time, however, the temperature will vary (especially if you are using an oil or water lens) and the likelihood increases that pressure and osmolarity will affect your specimen.

Another cheap and simple alternative is to use a commercially available plastic Petri dish with a coverslip in the base. This holds a larger quantity of medium than a slide and because the thermal contact with the stage is poor it will maintain its temperature for quite a while, as long as a dry lens is used. However, this same poor conductivity means large temperature gradients will arise if it is placed on a heated stage (Figure B.1), particularly if an immersion lens is used.

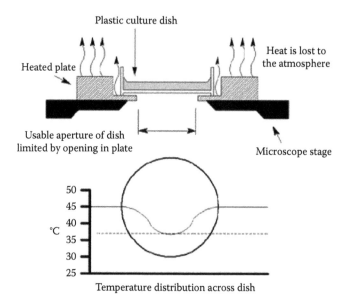

FIGURE B.1   There will be a large temperature gradient across a plastic culture dish used on a heated stage. (Courtesy of Bioptechs, Inc.)

Lenses are heat sinks. They have a large amount of metal in them and hence cause a huge temperature drain on a coverslip. Coverslips have been shown to move constantly over time when in contact with a nonheated objective. One cheap and very effective solution is to use a heating fan, which is directed across the whole microscope stage and objective. Once the microscope has reached a constant temperature thermal drift is no longer a problem. However, in the sealed slide model, cell metabolism affects the environment and pressure still influences your results. If you use an open chamber (of which there are many designs), the problem with a fan heating method is evaporation, exacerbating osmolarity effects. The heater will also make the microscope room quite warm if used for long periods.

A more sophisticated solution is to heat both the stage and the objective (Figure B.2). Various commercial solutions are available for this, but make sure you

FIGURE B.2   An objective lens heater. (Courtesy of Bioptechs, Inc.)

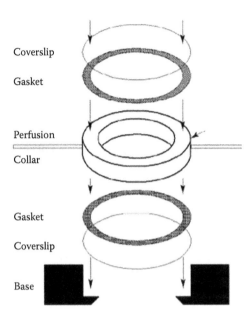

Coverslip

Gasket

Perfusion

Collar

Gasket

Coverslip

Base

FIGURE B.3 A simple perfusion stage. (Image, which has been modified, courtesy of Bioptechs, Inc.)

have a system with fast feedback and a small temperature variation, because a variation of as little as 0.5°C can cause changes within your cells. The choice between open and closed chambers is usually mandated by the experiment: Microinjection or micromanipulation demand an open chamber, and this kind of chamber is often the easier choice if you need to make changes to the chemical environment as part of the experiment. Thermal conductivity is important; it is possible to buy coated coverslips that conduct heat much better than plain glass.

The next step up in complexity is to use a chamber that has continual flow of solution over the sample (Figure B.3). The solution must be kept at a constant temperature. The simplest system for this is to have a solution flow across the chamber from a reservoir kept at constant temperature in a water bath. In a closed chamber, you must also then take precautions to avoid air bubbles.

How do you flow the solution without introducing pressure effects? You have two options: gravity flow or pump. Peristaltic pumps can introduce pulsing (and hence pressure effects), and it is important to minimize this. Nonperistaltic pumps are available but are more complex, and the more complex a system is, the more can go wrong. Gravity has the advantage of being simple, vibrationless, and free but as the medium is used up the head will vary and so the pressure and flow will change. This can be avoided, again with a slight increase in complexity, by including a constant-head arrangement with a subsidiary chamber. For dynamic physiological experiments you need to be able to change solutions over your sample quickly while keeping the sample volume small. Doing so involves a switching system just before the chamber. Again, several commercial models supply this.

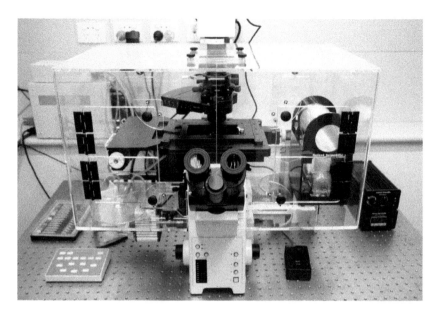

FIGURE B.4 Olympus Cell-R "whole microscope" environmental enclosure in the author's center.

In all of the preceding systems, we have assumed that the buffer system maintains the gas partial pressure. In the more advanced systems, however, the chambers can be enclosed to have a constant percentage of CO_2. Alternatively, one can progress to the next level of complexity, doing away with stage and objective heating, and enclose the whole microscope with both CO_2 and temperature control (Figure B.4). The problem with this approach is that it restricts access to the microscope controls. However, many automated microscopes are now available with fully electronic control, so that all operations can be carried out with a keypad or computer outside the enclosure. A moist atmosphere with high CO_2 is rather corrosive and conducive to fungal and bacterial growth, so thorough drying and cleaning between experiments is mandatory. In our lab we prefer to keep the temperature control on, both to dry everything out and in the interests of thermal stability. It is probably fair to say that the latest, large, controlled stages are now tending to swing the balance of opinion away from total enclosure.

Most live cell imaging systems are set up on inverted microscopes, though it is quite possible to use an upright microscope, which does have some advantages when you use dipping lenses. The upright approach is also often convenient for micromanipulators and microinjection. A fixed stage microscope, in which focusing is done by moving the objective, is normally used, as a conventional focusing microscope stage could not handle the weight.

LIGHT

Although it is generally true that the shorter the wavelength, the more damaging light is to cells, there are exceptions. In fact, different specimens may be sensitive to quite different wavelengths of light. For example, the movement of the fringes of cancer cells (Figure 16.2), an extremely important process in metastasis, has been shown to be sensitive to infrared irradiation. Prolonged, unprotected illumination causes the retraction of these protrusions.

Mercury and metal-halide lamps have an infrared-blocking filter built in, but the halogen lamps of conventional illuminators often do not and they produce a lot of IR. Therefore, if you are taking phase or DIC images, it is a good idea to add an IR-blocking filter, and a UV blocker as well, because there is also a UV component, albeit small, in the lamp output. For best contrast, phase images should always be acquired with a green interference filter rather than with white light (Chapter 2), but do not assume that this filter will do the blocking for you; unless you order them specially, optical filters are specified only for the visible spectrum. Continuous illumination (halogen or mercury) should be avoided for any experiment lasting more than a minute or so. In widefield microscopy, the best solution is a software-controlled shutter to cut off illumination between exposures, so that the cells are illuminated only when images are being captured.

MOVEMENT

Time magnifies all problems in live cell imaging. Even if we keep our cells alive and protect them from photo damage, there remains the problem of keeping them aligned with the field of view. Movement of medium in a chamber is liable to displace anything that is nonadherent, and thermal drift is almost certain to occur. Allowing everything to reach a constant temperature before starting the sequence of images can minimize thermal drift, and some commercial closed chambers use labyrinth systems to minimize turbulence in the perfusion medium (Figure B.5). Nevertheless, things move in any experiment lasting minutes or hours. The low-tech solution is to look every 5 minutes or so, recenter, then refocus. However, acquisition software that automatically maintains focus is available and effective. Even better are the systems that use a small infrared laser reflected from the coverslip to maintain focus. The Nikon Perfect Focus was the first in the field but most makers now offer something similar. If there are sufficiently distinctive features in the field, some software can also let the program track an object and keep it centered.

FINALLY

The last and most important thing to consider is controls. If you are imaging with an introduced fluorescent probe, you must run enough controls to make sure that the probe itself is not perturbing your specimen and hence influencing your results.

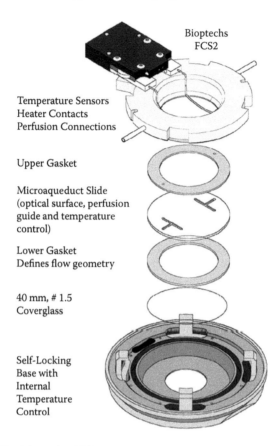

Bioptechs
FCS2

Temperature Sensors
Heater Contacts
Perfusion Connections

Upper Gasket

Microaqueduct Slide
(optical surface, perfusion
guide and temperature
control)

Lower Gasket
Defines flow geometry

40 mm, # 1.5
Coverglass

Self-Locking
Base with
Internal
Temperature
Control

FIGURE B.5 The Bioptechs FCS2, a closed-cell stage designed to optimize temperature uniformity and minimize turbulence in perfusion. (Courtesy of Bioptechs, Inc.)

Appendix C: Antibody Labeling of Plant and Animal Cells: Tips and Sample Schedules

Eleanor Kable and Teresa Dibbayawan

This appendix gives examples of antibody titrations, double-labeled specimens, and tips for handling antibodies.

The first experiment is an example of antibody titration. It is important to do both primary and secondary antibody titrations to determine the best dilution of each that is required to specifically label your sample. The first example is on a cultured monolayer of cells. The second example is how to carry out double labeling of plant cells and the importance of controls.

Always carry out a titration with each new batch of antibody that you buy. The antibody often comes with a recommended dilution, but this needs to be fine-tuned for each sample. Even if you have used a certain antibody before on the same samples, you should still do a titration, because antibodies vary from batch to batch and over time.

ANTIBODIES: TIPS ON HANDLING AND STORAGE

When the antibody arrives, it should be frozen in aliquots. Whenever possible, carry out a titration as soon as you receive the antibody. This allows you to determine the size of aliquots to be frozen. The amount frozen will be the concentrated stock that is needed for a typical experiment. It is not good practice to refreeze thawed antibodies or to store them too dilute.

Always check the supplier's instructions on how to store the antibody; some cannot be stored at $-80°C$, whereas others are more stable at this temperature. On receiving or defrosting the concentrated antibody, you should spin it down before diluting it. Use a microfuge on the top setting for 30 seconds.

When carrying out an experiment, always try to make up enough antibody for the experiment, because you cannot freeze the antibody once it has been diluted. Remember, the amount of antibody you have is usually quite small; if you use only as much antibody as is required for each experiment, the antibody will last for some time. If you are not careful, you will have enough to do only the initial test titrations.

PIPETTES: TIPS ON HANDLING

When using pipettes, consult the manufacturer's specifications. Over- and under-winding can cause inaccuracies. For example, a 20 µl pipette is often accurate only between 2 µl and 20 µl. Do not lay pipettes flat on the bench with the tips attached.

ANTIBODIES AND ANTIBODY TITRATIONS

In the following example, we use a 12-well slide (ICN Biomedicals Inc.) for both the primary and secondary antibody titrations. Before you proceed with the immunofluorescence protocol described here, you must calculate a suitable concentration range for each antibody over the 12-well slide, given the suggested range for each antibody.

In each case, you will need 10 µl per well (i.e., 20 µl for each concentration) because it is always best to do the titration in duplicate. You should make up a little extra, say 25 µl, to allow for inaccuracies in pipetting, and so forth.

You do not want to waste your antibody or be caught short midway through your serial dilution. Therefore, it *always* pays to calculate first how much stock antibody and buffer you require. Remember that you are doing a titration here, and require a minimum of 25 µl for each concentration.

Although you do the serial dilution starting at the most concentrated solution (the stock solution, in fact), you must begin the calculation for how much stock to start with at the other end: the most dilute.

Example

You are required to do a secondary antibody titration and have worked out that the six suitable antibody dilutions for a 2 × 6 well slide are as follows:

1. Control (buffer only)
2. 1 in 10 dilution
3. 1 in 30 dilution
4. 1 in 100 dilution
5. 1 in 300 dilution
6. 1 in 1000 dilution

How much stock should you start with? Remember that too much initial stock is expensive and wasteful, and so is running out of stock in the middle of a serial dilution.

The worksheet that follows is designed to help you through the calculations. As indicated by the calculate arrow at the far left of the worksheet, start at the top.

1. You need about 25 µl of the 1/1000 dilution (i.e., 10 µl in each of two wells, plus a bit left over for mistakes).
2. To get 25 µl of 1/1000 solution, you need to dilute some amount of the 1/300 solution with buffer. The amount you need is $300/1000 \times 25$ µl =

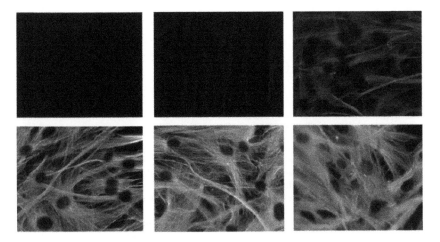

FIGURE C.1   These images show increasing concentration of α-tubulin. The secondary antibody used is FITC and the cells were HeLa cells. The top left-hand corner is the control. The antibody increases across the top of the picture and then goes to the bottom left-hand corner. The correct amount of antibody to use is the second to last concentration; the last concentration is becoming overstained and slightly blurry.

> 7.5 μl. In other words, 7.5 μl of 1/300 solution, diluted with 17.5 μl of buffer, will make up 25 μl of 1/1000 solution.
> 3. Therefore, you need to have at least 25 + 7.5 = 32.5 μl of 1/300 solution available when you get to this stage. The next dilution is a factor of 3, from 1/100 to 1/300. So make the amount we need a factor of 3, say 33 μl.
> 4. Repeat steps 2 and 3 until you get up to the amount of stock solution you need. Also, keep track of the amount of buffer you need at each step.

The suggested range for the primary antibody in the worked example is from 1/10 to 1/10,000. The secondary antibody is kept at a constant value. To titrate the secondary antibody, the primary is kept constant and the secondary is titrated. Figure C.1 shows the result of such a titration.

Primary antibody titration:

Primary antibody name:
Concentration used: Range (1/10–1/1000)

Secondary antibody name:
Concentration used: Constant value
Example calculation:

Well		Amount needed in wells (µL)	Amount transferred for dilution (µL)	Total needed of this concentration (round up to make maths easy) (µL)	Buffer needed (µL)
buffer	⚬⚬	25			25
least conc. 1/1000	⚬⚬	25	none	(25)	25 – 7.5 = 17.5
1/300	⚬⚬	25	$\frac{300}{1000} \times 25 = 7.5$	25 + 7.5 = 32.5 (next ratio is 1/3 so make this **33** µL)	33 – 11 = 22
1/100	⚬⚬	25	$\frac{100}{300} \times 33 = 11.0$	25 + 11 = 36 (next ratio is 3/10 so make this **40**)	40 – 12 = 28
1/30	⚬⚬	25	$\frac{30}{100} \times 40 = 12.0$	25 + 12 = 37 (next ratio is 1/3 so make this **39**)	39 – 13 = 26
most conc. 1/10	⚬⚬	25	$\frac{10}{30} \times 39 = 13.0$	25 + 13 = 38 (next ratio is 1/10 so make this **40**)	40 – 4 = 36
stock			$\frac{1}{10} \times 40 = 4.0$	stock needed: 4.0	Total buffer = 154.5 µL

(Left margin: Calculate — downward arrow. Right margin: Make up wells — upward arrow.)

IMMUNOFLUORESCENCE PROTOCOL

This experiment has been used over a wide range of cell lines, but you should do the titration on the cell line you are studying because the amount of antibody needed will vary from cell line to cell line. (The cells are grown on slides obtained from ICN.) The fixative you choose also depends on your cell line and on the antibody you are using.

METHOD

Add 10 µl to each well and *never* allow your slide to dry after you have added antibody to it.

1. Remove excess media from wells with cells by allowing media to drain off slide and wiping off excess with a tissue.
2. Wash in PBS in Coplin jar.   5 min
3. Fix using appropriate fixative by flooding slide with fixative.   30 min
4. Wash in PBS as in step 2.   2 × 5 min
5. Flood slides with blocking buffer: 3% BSA in PBS.   30 min
6. Incubate with primary antibody in PBS.   60 min
7. Use PBS wash bottles to rinse off excess antibody (low to high   2 × 5 min
   concentration, i.e., ICN label closest to wash bath) prior to placing them in
   Coplin jar as before.
8. Using folded tissue, dry off between wells. Be careful not to wipe off cells
   from wells.
9. Incubate in secondary antibody in PBS.   60 min
10. Wash in PBS as in step 7.   2 × 5 min
11. Stain with DAPI (1 µl/ml in PBS), a nuclear stain, in Coplin jar.   60 s
12. Wash in PBS in jar.   5 min
13. Mount in a suitable antifading agent (e.g., Citifluor in glycerol). Cut off
    end of a yellow tip and, using a 20 µl to 200 µl pipette, place 50 µl of
    Citifluor in the center of the multiwell slide. Cover with a coverslip,
    avoiding air bubbles where possible.
14. Seal with nail polish (or hot wax, if preferred).
15. Store slides in refrigerator in the dark using a suitable slide container.
    Leave slides overnight before examining to allow the antifade agent
    enough time to fully permeate the sample.

MULTIPLE LABELING AND DIFFERENT SAMPLES

It is important to remember that different tissues often require different specimen preparation protocols. In general, the preparation for plant material involves more steps than for animal tissue as a result of the presence of a thick and rigid wall in plant cells.

The aim of this exercise is twofold:

- To demonstrate the different preparation requirements for different samples
- To carry out a labeling protocol using more than one primary antibody

Immunolabeling with more than one antibody can be done either sequentially or simultaneously, depending on the antibodies. If the two primary antibodies have been raised in the same species, then sequential labeling is required. *Sequential labeling* involves one set of primary and secondary antibodies followed by the second set, with a fixation step in between. If, on the other hand, the primary antibodies are from different species, then you can apply *simultaneous labeling*, which mixes the two antibodies and adds them at the same time. This exercise involves simultaneous labeling of multispecies antibodies.

PLANT MATERIAL

Plant root tips are very good specimens for studying plant cell biology, particularly by immunofluorescence. The cells in the root tips do not have significant chlorophyll content; chlorophyll autofluorescence can mask and interfere with immunosignals. Plant root tips also have a large number of dividing cells, which are ideal for studying the different microtubule arrays in mitotic spindles. Because plant cells have a cellulose cell wall that forms a barrier to antibody entry, an enzymatic weakening or digestion of this barrier is required prior to antibody loading. The two antibodies that are suggested here are a mouse monoclonal anti-α-tubulin antibody directed against microtubules and a goat anti-clathrin antibody directed against coated vesicles and coated pits at the plasma membrane.

Protocol

1. Make up the fixative: 4% paraformaldehyde in PME buffer (50 mM PIPES, 1 mM $MgSO_4$, 5 mM EGTA). Use paraformaldehyde stock no older than 2 weeks.
2. Excise ~2 mm from the tip of onion seedlings (3 to 5 days old) in fixative solution and fix for 1 h at room temperature. Note that you can use forceps to nip off the root tip, avoiding root hairs.
3. Wash in PME buffer, 2 × 10 min (or 4 × 5 min) to remove all traces of fixative. Be sure to label your slide well (e.g., using a diamond pen).
4. Digest walls in enzyme mixture (1% cellulase plus 0.1% pectolyase) for 20 min.
5. While you wait, coat a 12-well slide with 0.1% PEI (polyethelenimine); this allows the cells to adhere to the slide. Dip a cotton bud in PEI solution (just enough to draw up a small amount) and paint each well to leave a thin film. One dip is more than enough for a single multiwell slide. Leave the slide to dry in a Petri dish before use.
6. Wash digested tips in PME buffer: 1 × 10 min, then 1 × 5 min. Remove excess buffer.
7. Use a plastic pipette to pick up root tips and place them on the multiwell slide. Fill each well of one slide with 2 to 3 tips, remove any excess buffer, and then squash tips under a coverslip by firmly tapping the coverslip with the tip of a pencil—not too hard but enough to squash the roots.
8. Carefully remove the coverslip without twisting or sliding. You should be left with a milky white liquid that contains the isolated root cells.
9. With a pair of fine forceps, squeeze and drag the larger pieces of debris around the wells before removing them. This encourages more cell release and adhesion to the glass surface. Remove excess debris.
10. Leave cells to air dry (or dry them in the incubator if you are in a hurry).
11. Rewet slide in PBS (40 μl/well) for 5 min.
12. Place in cold methanol (−20°C) for 8 min (20 μl /well).
13. Rehydrate in PBS for 2 min.
14. Block in blocking buffer: 1% BSA plus 0.02% Tween-20 in PBS (PBSTB), for 30 min by simply flooding the slide with blocking buffer.
15. Incubate in primary antibody diluted in PBSTB for 60 min at room temperature. You will be adding only 8 μl to10 μl of antibody solution per well. See the following diagram. With a 12-well multiwell slide, it is always

better to put controls in wells 6 and 12 because on some light microscopes the stage does not allow the objective to easily reach these wells.

16. Very carefully remove the antibody from each well *without* cross-contaminating other wells.
17. Wash in PBS, 2 × 10 min.
18. Incubate in secondary antibody diluted in PBSTB for 30 min at room temperature.
19. Wash in PBS, 2 × 5 min.
20. Stain with 1 μg/ml DAPI in PBS for 1 min and then wash briefly in PBS.
21. Mount in antifade agent (e.g., Citifluor) (~ 50 μl).
22. Seal the coverslip with nail polish.

Following is an example of some antibody combinations that have been used with this system.

Primary antibodies:
    1/20 Sigma goat whole antiserum against clathrin
    1/1000 Sigma mouse monoclonal antibody against α-tubulin

Secondary antibodies:
    1/50 Sigma rabbit anti-goat IgG Cy3 conjugate (RAG)
    1/200 Silenus sheep anti-mouse FITC conjugate or Molecular Probes Alexa
        488 rabbit anti-mouse (RAM)

Diagram Showing Position of Antibodies on Multiwell Slide

Control for Secondary 546	Anticlathrin Only	Double Labeling	Double Labeling	Anti-Tubulin Only	Control for Secondary 488
No 1° Ab	1/10 Anticlathrin	Anticlathrin + α-tubulin	Anticlathrin + α-tubulin	1/1000 α-tubulin	No 1° Ab
1/50 RAG-Cy3	1/50 RAG-Cy3	RAG-Cy3 + RAM-488	RAG-Cy3 + RAM-488	1/200 RAM-488	1/200 RAM-488
1	2	3	4	5	6
No 1° Ab No 2° Ab	Same as 2	Same as 3	Same as 4	Same as 5	No 1° Ab RAG-Cy3 + RAM-488
7	8	9	10	11	12
Control showing auto-fluorescence	Anti-clathrin only	Double labeling	Double labeling	Anti-tubulin only	Control for both secondary Abs

No fluorescence should be seen in the control wells. Only single label should be found in single label wells and double labeling in double label wells. Beware: If you do not dry between wells, you will have contamination from the other wells.

Appendix D: Image Processing with ImageJ

Nuno Moreno

INTRODUCTION

Acquiring pictures with a microscope can be the start of a painful process of enhancing relevant information, especially if you do not use the right tools. Besides enhancing information, image processing can be a matter of unveiling it. Hence, a formal perception of what is an image and which is the most suitable software might help you to extract all the information that an image can contain. For this reason, it is important to spend some time understanding the mechanics of image processing.

Two types of software exist: free and commercial. Each type has advantages and disadvantages, but for universal accessibility, we look at free software. This appendix does not intend to be a software manual; rather, it aims to impart some ideas of how to get some "hidden" information from microscopy pictures and to demonstrate some tips on how to present those pictures, concentrating on basic concepts.

ImageJ is public-domain image processing software, based on Java, that was inspired by NIH Image and developed by Wayne Rasband at the National Institutes of Health. Because it is written in a cross-platform language, ImageJ has been used repeatedly by many people, contributing to a tremendous database of software extensions, called *plug-ins*, which are freely available for download. For more detailed instructions, see Tony Collins's online manual at http://www.uhnresearch.ca/facilities/wcif/imagej or the ImageJ Web site at http://rsb.info.nih.gov/ij.

Because Java is a cross-platform language (or nearly so), it has some understandable limitations in dealing with hardware. But its big advantage is that many people are contributing to the image project. Even so, although some plug-ins let you control specific cameras or stages, for such purposes, commercial software still has no real free competitors.

DIFFERENT WINDOWS IN IMAGEJ

ImageJ has a few windows for specific purposes. The main window is ImageJ itself, which contains menus, functions, and plug-ins that you can see hierarchically by selecting *Plugins → Utilities → Control Panel*. The second most important window is the image window, which can contain a single picture or a stack. (In the case of a stack, a scroll bar appears at the bottom of the window.)

Other windows include:

- The Results window, which displays measurement information.
- The Log, which receives output information such as errors or other system information.
- The Macro Recorder and Editing windows, which record all the steps you make for possible process automation later.

IMAGE LEVELS

As explained in Chapter 6, an image is a matrix of small dots called *pixels* (for a 3D data set these are called *voxels*). In the case of an 8-bit grayscale image, pixels have intensities distributed from 0 to 255. Changing contrast and brightness is probably the first basic operation in image processing; in ImageJ, you can do this by selecting *Image → Adjust → Brightness and Contrast*. If the intensities that we see on the screen are correlated with real information by $I = i \times c + b$, by increasing brightness we are in fact increasing b, meaning that the output (vertical axis or I in this case) will be higher. In the case of contrast, we change the slope; increasing c means that the low-intensity pixels will value even less and high-intensity pixels will value more. However, there are two more sliders in the window: *maximum* and *minimum*. These sliders represent another (and probably easier) way to get the same results. If our image has only 200 gray intensities, then we want to change the maximum to 200 instead of 255 in order to have pixels on the screen change from black to white. This principle can also be used for getting rid of the background by changing the minimum value to be displayed (Figure D.1B).

A simple way to change the response function of the monitor is to pass from a linear to a nonlinear response, emphasizing the lower- or higher-intensity pixels. In the case of Figure D.1C, there was a change to a square root instead of a line, meaning that our intensity function is now $I = c \times \sqrt{i} + b$. In this case, when we emphasize the lower-intensity pixels, the whole picture looks lighter. After a nonlinear process like this, intensity-based quantification of the image is no longer possible. This function can be found on *Process → Math → Gamma* with, in the case of square root, 0.5 as parameter (Figure D.1C).

COLORS AND LOOK-UP TABLES

Chapter 6 mentions that all the sensitive devices detect photons not their energy. This means that most of the colors that we see in fluorescence microscopy pictures are coming from black-and-white cameras, but if we are looking at green fluorescent protein (GFP), it is more comfortable for our brain to see it as green. Without changing the way a computer deals with a picture—or rather, without changing from 8 bits to 24 bits (8 per channel)—it is possible to change colors by selecting a different look-up table (LUT). By doing this, we still have the same 8-bit picture, but it is displayed in a different way (Figure D.2A).

If initially we add an LUT just for viewing comfort, there is an important secondary benefit. While it is true that our eyes better distinguish changes in gray than in

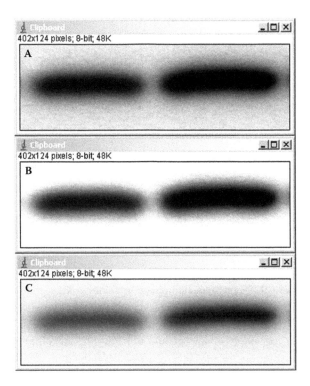

FIGURE D.1 Adjust Brightness and Contrast window. Usually, the best setting is to adjust the output level in order to have the maximum and minimum intensities displayed on the screen as white and black, respectively. (A) Original picture. (B) Change in maximum and minimum to set background to white and minimum to black. (C) Changing gamma effect changes the output levels in a nonlinear way, for example, to enhance low-level features.

other colors, it is very useful if we change from intensity to a color contrast, since we can better perceive differences in color than differences in intensities (Figure D.2B). This is especially true in pictures with a small dynamic range, as in the case of ratio imaging. ImageJ has several tools for managing color tables. The color table in the second image of Figure D.2 was taken from another picture by using *Image → Color → Edit LUT* and then saving it. You can then use this look-up table whenever you want.

Because our sample usually has more than one color, we might want to show the colors all together. If you have two channels, for example, you can merge them by using *Image → Color → RGB Merge*, which lets you select which channel goes with each color.

SIZE CALIBRATION

Size information is very important when you display pictures. This can be obtained by making a spatial calibration of your microscope or camera. Please note that any optical or camera change in the system will affect this calibration, such as changing

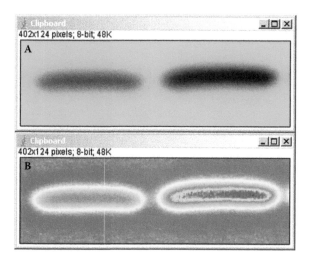

FIGURE D.2 You can attribute different color palettes to 8- and 16-bit pictures. (A) In a green look-up table (LUT), the maximum green brightness corresponds to the maximum pixel intensity and the black to zero. (B) Other LUTs can be used to enhance small differences in intensity by changing color rather than intensity.

the distance between the C-mount adaptor and camera, changing the camera to a higher resolution, binning, or adding a tube lens magnification.

In practice, you must take a picture of a microruler (or any micrometric shape with a well-known size), and then use the software to make the correspondence to the number of pixels by making a straight-line ruler image (Figure D.3). By making the calibration global, you ensure that it is valid for other opened pictures. Remember

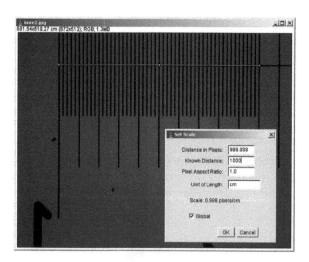

FIGURE D.3 A microscope can be used as a micrometer. However, you must calibrate it for the first time, which can be done by taking a picture of a microruler (stage micrometer). In ImageJ, you can then attribute the distance to the measured line in microns.

that this is a linear variation; if in this case we have 1.014 μm/pixel or 0.986 pixels/μm on a 10× objective, we need to divide by 6 in the case of a 60× objective, ending with a pixel size of 0.169 μm.

IMAGE MATH

Images can be considered as a matrix because they contain a coordinate and a value, so it is simple to perform operations with them. The most common operations are probably the division used for ratio imaging (Figure D.4) and the subtraction for background correction. Other operations can be done as well, either mathematical or bitwise, but these are less common in biological images.

Ratio imaging can be used for intensity quantification when we want to minimize the artifacts due to heterogeneous dye concentrations. In the present example (Figure D.4), the dye is the calcium indicator Indo-1, which has one absorption and two emission peaks. By dividing one channel that is calcium dependent (Figure D.4A) by the other one that has the information about dye distribution (Figure D.4B), we can then have a picture with more accurate data about the ion distribution. A simple way to do this in ImageJ is to multiply one of the channels, usually the one that intensity increases with the increase of ion concentration, by 255 and then divide by the second channel. Doing so requires these steps:

1. Convert first image to 16 bits: *Image → Type → 16 bits*
2. Convert second image to 16 bits: *Image → Type → 16 bits*
3. Multiply by a factor: *Process → Math → Multiply*
4. Adjust the image levels: *Image → Adjust → Brightness/Contrast*
5. If desired, apply a median filter: *Process → Noise → Despeckle*
6. Divide the two pictures: *Process → Image Calculator → Divide*
7. Change the color table: *Image → Lookup Tables → …*
8. Display LUT: *Analyze → Tools → Calibration Bar*

Figure D.4 shows the result.

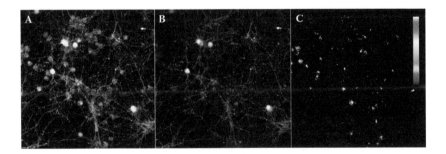

FIGURE D.4 Ratio image of primary cerebellar granule neuron cells stimulated with quinolinic acid. Cells were incubated with Indo-1, a ratiometric dye. Images A and B come from the dependent and independent channels, respectively. Image C comes from the division of image A by B.

QUANTIFICATION

Image quantification can be applied to different properties, which can be classified by intensity, size, shape, motion characteristics, particle amounts, and so forth. However, bear in mind you have to quantify at acquisition time. Doing so avoids problems like overexposure, background correction, standard experiment settings, and so forth. In intensity quantification, another thing to consider is the detector linearity. Black-and-white CCDs are usually pretty linear and so are PMTs in most of their working voltage. Independently of the detector used, relative intensity quantification is easier than absolute measurement; this is where we will focus our attention.

In the case of intensity measurement, the first step is to measure the background. For fluorescence, you take this measurement in two steps:

1. Measure the nonuniformity of illumination distribution by taking a picture of an homogeneous sample. (There are commercial solutions for this and acrylic slides of a specific color.)
2. Measure the negative control of the sample per se for checking, as an example, whether antibodies are nonspecific.

Next, subtract these values from the original picture by using *Process → Image Calculator*; select the subtract function. You can now measure these pictures in terms of intensity, comparing intensity not only on the same picture but also to others taken under the same conditions. (Keep in mind that over the lifetime of a bulb or laser intensity diminishes, and misalignments due to intrinsic or extrinsic factors can occur.) Consider taking pictures as close as possible in time, preferably on the same day.

In all such measurements you need to play with *regions of interest* (ROIs). In ImageJ, regions depend foremost on intensity. By thresholding your picture using *Image → Adjust → Threshold*, you can select an area individually by using the wand function. By using *Analyze → Measure*, you then have access to some region quantification measurements. In this case, we can see mean gray intensity, standard deviation of gray intensities in the region, and so on (Figure D.5). For selecting the parameters that you want to access, go to *Analyze → Set Measurements*. If you have several regions to analyze, consider using *Process → Analyze Particles*.

STACKS AND 3D REPRESENTATION

Modern acquisition systems (i.e., microscope, camera, and software) can collect not only XY images but also in Z, time and wavelength. To deal with multidimensional acquisition, we use a pile of ordered pictures, or *stack*. Like many other image-processing programs, ImageJ can work with stacks.

A common example of a stack is a time-lapse experiment. In the next sequence we will combine some operations showing some possible applications. By scrolling the stack through time, we see almost no difference between frames except a very small bacterium with very low contrast moving around (Figure D.6A). By subtracting the

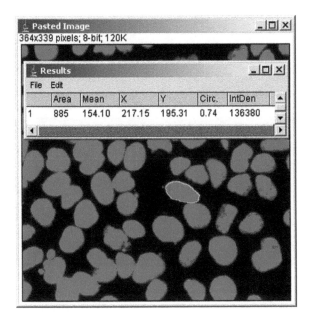

FIGURE D.5   Intensity quantification by using threshold. Because the image is not spatially calibrated, the area will be in pixels. Mean is the average gray value, XY are the centroid coordinates, circularity measures the approach to a circle, and integrated density is the integral of pixel intensity or the mean value times the area.

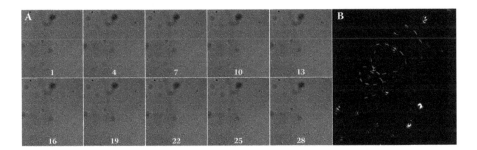

FIGURE D.6   (A) In the montage we see almost no differences among pictures in the stack. (B) In order to enhance some differences, a background subtraction was done.

first image, which contains immobilized bacteria, from the rest of stack, we clean up the unwanted information. To do this, slide the stack into the first position and then choose *Image → Duplicate*, without selecting for the entire stack. Next, choose *Process → Image Calculator* and make a new window with the subtraction of the stack from the duplicated image. For the new stack, make a maximum intensity projection using *Image → Stacks → Z project*. Adjust the maximum level until you reach something similar to Figure D.5B. To remove some noise, use a despeckle filter, which you can find by selecting *Process → Noise → Despeckle* (see Chapter 10 for convolution filters).

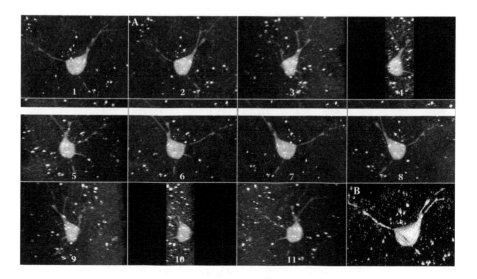

FIGURE D.7 (A) Neuron 3D projections using the right distance between frames with interpolation, rotation interval of 30°. (B) Volume reconstruction using VolumeJ, a plug-in for 3D rendering.

As mentioned in Chapter 11, modern microscopes also acquire in Z, so somehow we need to show the third dimension. The most basic way is to make an extended focus using a maximum intensity projection, as previously described. However, if we use the same principle but make the projection by using different viewing angles, we can build a new data set that contains data rotating around a specified axis (Figure D.7A). To do this, select *Image → Stacks → 3D Project*. The other way to show information in a 2D space is stereo imaging; however, there is no direct way to do that in ImageJ, and the need for special glasses makes this process impractical. Volume rendering can be an interesting method for 3D representation (Figure D.7B), but speed is a big drawback if you use ImageJ for that purpose.

FFT AND IMAGE PROCESSING

As Chapter 10 showed, an image can be represented in spatial or frequency domain. In spatial domain, intensities are distributed in a way our brain understands; in frequency domain, it is impossible to understand intensities unless they come from very simple patterns. Despite having other applications, such as pattern recognition and image compression, frequency domain image processing in microscopy is restricted mainly to filtering, but does it in a very elegant way. Figure D.8A shows an image of a growing pollen tube with a repetitive pattern coming from the monitor. After erasing those identified frequencies using the fast Fourier transform (FFT) through *Process → FFT → FFT* and the painting tool (Figure D.8A, B) we can then go back to the processed picture by using *Process → FFT → Inverse FFT*, resulting in a picture without any bright stripes (Figure D.8C). If you want to avoid all this trouble,

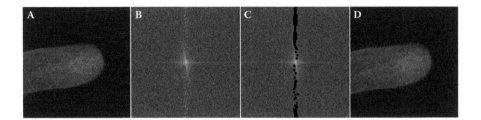

FIGURE D.8　(A) In this growing pollen tube, we can see some stripes coming from monitor scan frequency that were caught by the PMT. (B) Once they are repetitive, it is easy to discriminate them in frequency domain. (C) These frequencies are then erased from the picture. (D) An inverse Fourier transform then eliminates those stripes almost completely.

you might try the band-pass filter on the FFT menu, where you can select whether you want to remove horizontal stripes or, vertical stripes, among other parameters.

MACRO LANGUAGE IN IMAGEJ

In biological samples, one often needs to carry out repetitive processes that can be more or less automated, depending on human selection criteria. Most of the good commercial packages let you access built-in functions for use with flow controls, conditions, and so forth. For example, Image-Pro from Media Cybernetics has Visual Basic for Applications built in; MetaMorph from Molecular Devices has a different philosophy, using a higher-level language called Journals. ImageJ offers a macro language with a syntax very similar to Java, or you can create a plug-in in Java (doing so is complicated and is not covered in this appendix).

Despite looking scary at first for people not used to computer programming, the macro language is very simple to use, especially with the record macro function in *Plugins → Macros → Record*, which writes in a window all the commands that you are using in the program. Next, click Create Macro, and you have code without understanding anything about programming. However, this code will not be very useful if you do not add any kind of loop or selection conditions for computing a large number of images with different conditions.

In the next example, we will try to repeat the process that originated Figure D.5B but using 10 different stacks. If we had to do this by hand, we would need almost 20 clicks times 10 stacks, if there were no mistakes. If file names are saved conveniently, however, it is easy to access each stack and save the result in the same folder with the name *result<n>.tif*:

```
// set up our loop
for (i=0;i<10;i++){
     //open our files
     open("D:\\cd2\\demo pics\\dvhlooping" +i + ".tif");
     run("Select All");
     run("Copy");
     //subtract the first frame from the stack
```

```
    newImage("Untitled", "8-bit Clipboard", 400, 400, 1);
    imageCalculator("Subtract create stack","dvhlooping.
tif","Clipboard");
    run("Delete Slice");
    //project the processed stack
    run("Z Project… ", "start=1 stop=29 projection=[Max
Intensity] ");
// normalize the contrast
    run("Enhance Contrast", "saturated=0.5 normalize");
    run("8-bit");
//save the image
    saveAs("Tiff", "D:\\cd2\\demo pics\\result" + i +".tif");
    close();
// go round the loop again
+
}
```

If all the files to be processed were exclusively in a folder, we would need a slightly different code for looping all files in a directory by using functions like getFileLis(), which would return an array with all the file names from a certain directory.

More sophisticated automation than what is shown in the example can be done, such as writing selectively in the Results window, plotting simple graphics, managing ROIs, and so on. Notwithstanding, the limitations of ImageJ macro language are easily reached, especially where variables are concerned. It is not possible, for example, to have multidimensional arrays and pass an array as a parameter, which is very convenient when dealing with pictures. You can go further, learn Java, and make a plug-in instead of a macro, but in doing so you would actually be extending the program.

Index

A